BRUNEL'S SHIPS AND BOATS

Helen Doe

AMBERLEY

To my mother 'Gee', who inspired my love of maritime history.

First published 2018

Amberley Publishing
The Hill, Stroud
Gloucestershire, GL5 4EP

www.amberley-books.com

British Library Cataloguing in Publication Data.
A catalogue record for this book is available from the British Library.

ISBN 978 1 4456 8364 5 (print)
ISBN 978 1 4456 8365 2 (ebook)

Origination by Amberley Publishing.
Printed in Great Britain.

Contents

Acknowledgements   4

An Early Interest in Ships   5

Industrial Design   9

Across the Atlantic   11

Passenger Requirements   21

A Sister Ship   28

Propellers and the Royal Navy   31

Launch of the *Great Britain*   34

The *Great Britain* in New York   38

Disaster   41

The Australia Years   44

1852: The First Australia Run   46

Crimean War Service   48

Passengers to Australia   51

A Cargo Ship   55

Brunel's Shipping Projects after 1850   57

The Great Ship   61

Gunboats   70

The Great Ship Launch   72

The Atlantic Cable   81

The *Great Britain* in the Twentieth Century   88

Brunel's Impact on the Maritime World   91

Time Line   93

Selected Further Reading   95

Image Acknowledgements   96

Acknowledgements

A short book such as this cannot do full justice to Brunel's shipping ventures and its intention is purely to give a narrative overview of his work. With one notable exception, books on Brunel's maritime projects have tended to focus on individual ships. This can make them seem to be adjuncts to his big engineering projects, such as the Great Western Railway and the great bridges. By bringing the vessels together in this publication, and by including some rather more humble craft, it provides a picture of Brunel's continual commitment to naval architecture, as well as the advancement of ship design and marine engines throughout his life.

There is no space for footnotes, but I have indicated in the text the origin of key quotes. The sources have mainly been the many excellent articles and books on his ships and key books are listed at the end. Some original sources have also been consulted, such as the University of Bristol Brunel Collection held at the Brunel Institute of the SS Great Britain, contemporary newspapers, and ship and company registration documents at the National Archives. There is potential for much more research to be done in several areas.

Thank you to Andy King of MShed who provided extra information. At the SS Great Britain Trust, I am indebted to Kate Ashton, Nick Booth and Mollie Bowen for their considerable help with images and documents. Joanna Thomas, Maritime Curator at the Trust saved me from several errors and provided valuable input. Any other errors are all mine. Matthew Tanner, chief executive of the SS Great Britain Trust has, as always, been an enthusiastic sponsor. One sadness is that my good friend, and proud supporter of the *Great Britain*, Tony Dickens, did not live to see this next book although we had a great conversation about it before he died. My patient, long-suffering, supporter is my husband, Michael.

An Early Interest in Ships

Isambard Kingdom Brunel, like many of his contemporaries, did not see occupational boundaries. As an engineer his interest was in resolving problems, testing new technology and pushing boundaries, whether that would be in bridges, tunnels, railways, docks or ships. We know him well for his wonderful work on all these constructions and in shipping he is best known for the magnificent iron steamship the *Great Britain*, now at Bristol, and for his unlucky leviathan, the *Great Eastern*. Less well known is his first ship, the *Great Western*. It would be a mistake to think that he was the creator of just these three ships. Vessels of all sizes exerted a fascination for Brunel from an early age and were a strong theme through his career as this book will demonstrate.

By the beginning of the nineteenth century, wooden sailing ships had been the standard conveyance across the seas for cargoes and passengers for centuries and had largely reached their limits in construction. Some small improvements could be made with hull design, internal bracing and rigging, but it would take new materials and new forms of propulsion to make the great breakthrough. Brunel was born in April 1806 into this

Great Western at sea. (New York Public Library)

Great Britain approaching Cape Town, painted by John A. Wilson. (Clive Richards Collection)

Great Eastern off Gibraltar. (Clive Richards Collection)

new age of exciting possibilities in sea transport and he delighted in the engineering opportunities and challenges they posed.

Visualise a time when railways were in their infancy, the Liverpool to Manchester railway was the first to open in 1830, and shipping crowded the many ports and harbours around Britain and Ireland. Bristol, where Brunel would build two of his large ships, was then still a nationally significant port. In the 1830s it had 316 ships (49,535 tons) registered. Liverpool was still establishing itself and had 805 registered ships (161,780 tons) and was smaller than Newcastle with 987 ships (202,375 tons). London still dominated with 2,663 ships (572,835 tons). The merchant fleet was growing as Britain's trade expanded and this is reflected by the expansion of the number of men and apprentices employed on ships from 100,509 in 1840 to 170,628 just three years later. It is also a reflection of the steam revolution in which Brunel played such a large part since steam ships were larger and required more crew.

From a young age Isambard Brunel played with boats. While at school in Hove, from where he had a vantage point of the shipping in the English Channel, he wrote home in 1820 with the news that 'I have been making half a dozen boats lately, till I've worn my hands to pieces.' The type of boats is unknown, but there is a good case to suggest that these were steamboats. While steam power had been in industrial use elsewhere for a while, marine steam was just emerging and was rapidly making its mark. From the first British experiments with the *Charlotte Dundas* in 1803 to the *Comet* in 1812 entrepreneurs saw the benefits of marine steam. By 1815, when Isambard Brunel was just nine years old, there were several steamboats in operation, mainly in rivers, but the *Thames* steamboat had already proved itself by steaming from Glasgow to London. In 1817, he witnessed his father's experiments with steam shipping on the River Thames. Marc Brunel was keen for the navy to see the advantages of steamships, which could not only tow warships when there was lack of wind but would also be able to assist warships in distress. The young Brunel watched with his two siblings from London Bridge as his father demonstrated a steamboat in the towing trial on the Thames, hiring the *Regent*, built in 1814 with engines by his good friend, Maudslay. Just three years later, Isambard at the age of fourteen is making not one but dozens of model boats, perhaps trying to assist his father's ideas. Marc, who had a strong influence on his bright young son, continued his interest and took out patents on marine engines in 1822. Although, when asked to act as consulting engineer for an early experiment with steamships to the West Indies, Marc replied, 'As steam cannot do for distant navigation, I cannot take part in any scheme.' In view of the technology of the time, he was quite right, and it would be his son who would achieve this feat.

Isambard Kingdom Brunel came of age at twenty-one in April 1827 and in that year he noted his thoughts in a private diary. At this stage in his career he was his father's right-hand man on the great project of the Thames Tunnel. On 13 October in quiet reflection he wrote about his ambition. It was not he said, 'a mere wish to be rich', but he had a dream or 'castle in the air' of leading a fleet of naval steam vessels and taking 'some island or fortified town' such as Algiers or 'something in that style'. He and his father at the time were working on an alternative to steam called a 'gaz engine'. Brunel wanted to 'build a splendid manufactory for gaz engines, the yard for building the boats for them. And then at last be rich and have a house built.' He was to achieve most of these dreams. While the experiments on gaz engines came to nothing, his work did lead to many fine ships and even the yards and manufacturing places for building the marine engines.

Above: Sir Marc and
Lady Sophia Brunel.
(Clive Richards Collection)

Left: Isambard Kingdom
Brunel. (Clive Richards
Collection)

Industrial Design

Isambard Brunel's chance to move out of the shadow of his distinguished father came in 1831. Brunel heard about a competition to design a bridge across the dramatic Clifton Gorge in Bristol. When in Bristol, he was also asked to assist with the problems of the Bristol dock. The floating harbour with its double locks into the Avon River was in the heart of Bristol but it tended to silt up. Brunel examined the harbour and its water supply and in his report to the Dock Company in 1832, together with a range of other recommendations, he suggested a method of clearing the silt using a small boat with an iron scraper to drag the mud, 'by means of flat bottomed barges with large hoes or scrapers, strongly attached to them and suspended in such a manner as to be capable of being raised or lowered according to the depth of water or the consistency of the mud'. Bristol had been one of the very earliest ports in Britain to adopt steam technology for its uses in towing ships up and down the winding Avon River and it was a natural step to power the drag boat with a small steam engine.

The drag boat was designed in conjunction with John McLean of Bristol, who was probably employed by Lunnel & Co. Brunel's involvement with the Bristol Docks Company began in 1832 and in his correspondence with directors from 1832 to 1839, the drag boat occurs in frequent references and it is clear he took a personal interest in this small, industrial and unromantic craft. In May 1834 he wrote concerning repairs to the boat and the next month in a letter to an unnamed correspondent, perhaps the manufacturers, he again discussed various defects in the drag boat ordered for the Bristol

Drag boat BD6 paddle.
(Bristol Museums and Art
Gallery)

Dock Company and asked for them to be repaired before delivery. His interest was not just as an observer. On 22 June 1834 he described a visit to Bristol docks when he personally used the drag boat to clear the mud and scour the sluices.

Brunel's plan was for two vessels scraping a hoe-full of mud into the middle of the basin from which it could be swept away by more conventional scouring. In practice, the first drag boat operated on its own, warping itself across the dock basin by hawsers attached to bollards on the opposite side, and this arrangement worked well for ten years from 1834; it was then replaced by a slightly larger boat (eventually called BD6). By 1842 Brunel was still writing to the Bristol Dock Company concerned that some of his recommendations had been abandoned. Trade was increasing nationally; many ports were under pressure and Brunel had a ship waiting to launch into the floating harbour. He suggested deepening the basins, increasing water flow through the float, and he proposed a new drag boat and gave an estimate of cost. This vessel remained in service in Bristol until 1961 and parts of it survive in the collection of MShed, Bristol's industrial museum, as well as original plans and photographs of the vessel from the 1950s and '60s.

The SS Great Britain Trust now owns *Bertha*, believed to be the oldest 'operational' steam driven vessel in Britain, and possibly the world according to the National Historic Ship Register. The reason for the acquisition is that this was designed and built in Bristol around 1840 for use at Bridgwater, to the same design as Brunel's original. *Bertha*, like her Bristol predecessor, was still in operation in Bridgwater in the 1950s.

When Brunel was testing his first small steam vessel, the humble drag boat, he had his first major appointment when he became engineer to the proposed Great Western Railway in 1833. This was to absorb an enormous amount of his time over the next few years as he drove and rode energetically backwards and forwards between London and Bristol, negotiating, cajoling, demanding and managing every detail. Yet, he still managed to drive his ship projects with the same amount of enthusiasm.

Bertha in action in Bridgwater Dock. (Richard Witcombe/SS Great Britain Trust)

Across the Atlantic

It was in 1835, during a board meeting of the railway directors, that much was made of the fact that the new proposed railway line would be the longest in England and it is Brunel who is attributed with saying, 'Why not make it longer and take it to New York?' At the time this was a daring idea since steamships were largely restricted to coastal and river work, mainly because their engines were so inefficient that there was a need for frequent refuelling. Since the *Comet* in 1812, there had been much experimentation and gradually the distances were getting longer. But the Atlantic was thought by many to be a step too far for existing technology. One eminent scientist of the day, Dr Dionysius Lardner, was particularly vocal and as a well-known academic his views carried more weight than that of an untried and untested engineer. Two North American vessels, *Savannah* and *Royal William*, had crossed the Atlantic using both sail and steam, but they only attempted the eastward crossing where they could get the best advantage from prevailing winds. Brunel's idea was to set up not just a test journey but a regular line crossing, linking New York with Bristol.

Fortunately, there were others who believed in his vision and a company called the Great Western Steamship Company was established. A building committee made up of Brunel, Christopher Claxton, Thomas Guppy and the Bristol-based shipbuilder William Patterson met to decide on the best design for the steamship. Some of the committee members went on a tour of Britain to examine the current state of manufacturing and reported their findings. The ship was built as a wooden paddle steamer by William Patterson very much to the design of Brunel and his colleagues.

The initial plan of the new company was to build two 1,200-ton ships each with 300 hp engines and each ship would cost about £35,000. A prospectus was issued, and the first meeting of the new company took place on 3 March 1836. Subsequently Brunel recommended that the company should build one ship of 1,400 tons with a 400 hp engine, adding the prophetic words 'because the larger the ship the more efficient it is'.

It was important to the directors that the ship should be Bristol-made since many of the shareholders were Bristol or Bath-based. As much as possible of the ship and its internal fittings was commissioned locally. The engines however were a different matter as the whole success of the expensive project depended on their efficiency. Bristol did not have engineering facilities of the right standard in what was still an emerging profession and the contract went to Maudslay, Son & Field, who were long-time collaborators with

Above left: Thomas Guppy. *(A Short History of the Great Western*, published 1938)

Above right: William Patterson. *(A Short History of the Great Western*, published 1938)

Below: Bristol Harbour, showing sites for *Great Western* and *Great Britain*. (Denis Griffiths)

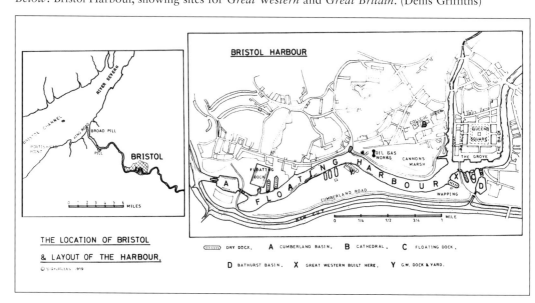

Brunel and his father. Maudslay was one of the world's leading manufacturers of marine engines and Brunel and his father had known the firm for many years. Brunel paid close and personal attention to the progress of the engine building, knowing that the success or failure of the whole grand scheme was wholly dependent on them. In a letter from Brunel to Christopher Claxton, the Managing Director of the Great Western Steamship Company, he reported on his visit to Maudslay to inspect the engines. 'Great exertions were apparent' and he had hopes that…

…we might affect that most important object of performing a voyage across the Atlantic this autumn. With respect to the quality of the work as far as it is completed and the very judicious modification of the difficult parts to unite the circumstances of the great increase of dimensions I never felt any anxiety, but I think it right to say that they appear to me such as to ensure a degree of power and economy and perfection in the working which promise every advantage we can wish for in the future competition with other vessels and particularly that of certainty even in our first voyage.

While the shipbuilding was in progress others matters such as crewing needed attention. An important appointment was the master of the vessel. Upon his shoulders rested the responsibility for the safety of the ship the crew and the passengers. He had to manage not just a conventional sailing vessel but also a department of engineers with new technology and the department of stewards and waiters and cooks to look after the special passengers. The man they chose was Captain Hosken, an ex-naval officer. He had

Portrait of Captain Hosken as a naval officer. (Hosken family/SS Great Britain Trust)

limited experience of steamships, which was not unusual for the time, steamships being still much in their infancy, but his first officer was Captain Bernard Matthews, who had plenty of experience in steam but less experience, as it would transpire, in that crucial skill of dealing with passengers.

The ship was launched on 22 July 1837 and drenched in Madeira by Mrs Miles, the wife of the local Member of Parliament, in front of a massive crowd of cheering onlookers. Once launched the ship then proceeded under sail to London, where the engines and paddle wheels were to be fitted at Maudslay's works at Vauxhall.

The fitting out of the ship in terms of its décor and furnishings, even down to its crockery and cutlery, was to be of the very best. Steam vessels had to compete with sailing ships and steam vessels were noted for the 'splendour of the saloon'. The ship was designed to carry only one class of passenger, superior class, and no effort was spared to provide them with the best. Noted artists of the day were brought in to decorate the saloon and the cabins.

While in London the novelty and the scale of the ship attracted vast and curious crowds. At last on 31 March 1838, after many engine trials and when the last workmen had finished laying carpets and decorating walls, she set out under steam to proceed down the English Channel to return to Bristol. Crowds cheered her off from the shore and a large party joined the vessel. Both Brunel and his father, Sir Marc Brunel, were on board and Sir Marc and other dignitaries disembarked at Gravesend. Shortly afterwards there was a calamity when some felt caught fire around the base of a funnel. Hosken and the pilot rapidly steered the ship to a mud flat. Fire on board any ship is

Launch of *Great Western*. (Denis Griffiths)

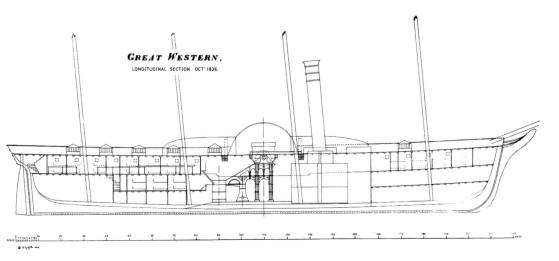

Great Western longitude, 1836. (Denis Griffiths)

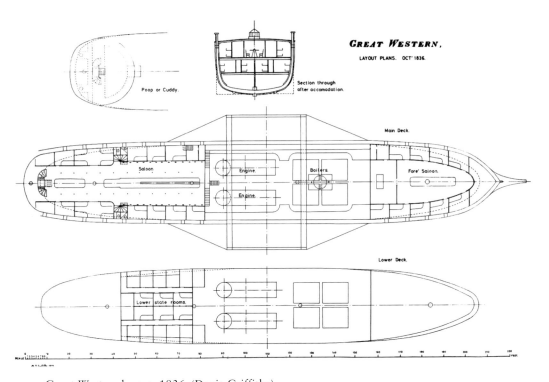

Great Western layout, 1836. (Denis Griffiths)

a frightening prospect, even more so on a wooden vessel, and in the ensuing confusion Pearne, the chief engineer, and Captain Claxton rushed to man pumps and a small manual fire engine to put out the fire. Brunel also hurried to assist and clambered down an iron ladder, but had a very bad fall and was rescued by a quick-thinking Claxton.

Several firemen on board also panicked and jumped into a small boat and rowed for the shore. Brunel, who was thought to be seriously injured, was taken ashore. Although he was severely concussed and bruised he eventually made a full recovery, but he was not to travel on his great project.

The ship proceeded without Brunel to Bristol and moored in the mouth of the Avon River to finish the final fitting out of furniture, load stores and the rest of the crew and await the very first passengers and cargo. The company had promoted the first passage extensively and the advertisements for this first experimental journey were reassuring. The ship was described as strongly built and copper fastened, with engines of the 'very best construction, by Maudslay, Son and Field and expressly adapted for the Bristol and New York station'. This was to emphasise that this ship was not merely some experiment but had been specifically designed as an Atlantic liner. After explaining how to book a passage, there was an additional last-minute thought added in italics at the bottom of the advertisement to reassure potentially cautious passengers: 'This ship has coal stowage for 25 days constant steaming, and therefore will not require to touch at Cork for coal.'

Until then there had been a reasonable number of bookings, nearly fifty passengers, for the ship on its very first passage to America, but perhaps the dire predictions of the experts and the fire incident put off some of the passengers and in the event just seven people stepped on board. These six men and one intrepid woman were all outnumbered by the crew including the waiting staff and steward. The passengers comprised two American merchants, Foster and Welman, both age thirty-one, a British Army officer, Colonel Vernon Graham, age fifty-five, and two other merchants: John Gordon, age twenty-five, and the wonderfully named Cornelius Birch Bagster from Canada, age twenty-five. Charles Tate or Tait, aged twenty-five, was a Yorkshireman, a civil engineer and it is quite possible that he was Brunel's protégé who had represented him at earlier celebrations of the raising of the ship's stern the previous year. While it is doubtful if Brunel himself had planned to travel with the ship it is very likely that he would want one of his men on board to check on progress and observe. So, it was a mixture of those who had business to conduct and those who were there for the grand experiment. The odd one out was Miss Eliza Cross, single, aged twenty and travelling alone. She was described as a 'proprietress of stay, straw and millinery warerooms in Bristol.' A brief mention of her death in 1893 described her as the niece of Captain Hosken. While this gave her added protection as a lone woman traveller it was a brave venture for a single woman.

The intrepid passengers made their way downriver from Bristol on a small steamer to join the ship. At 8 a.m. the next morning the engines were fired and the adventure began. The journal of one of the Americans, Foster, shows that all on board were well aware of their potential place in history as he carefully noted each moment on board for future publication.

As the *Great Western* headed into the Atlantic she met an early challenge of a severe storm but weathered it well, the only damage being the denting of the Neptune figurehead, which lost its trident. The main anxiety was in the engine room where Pearne, the Chief Engineer, kept a close watch on every aspect and the firemen toiled day and night feeding the fires with coals and removing the debris. For some of these men this was their first time at sea and there were discipline issues to contend with for the captain. When not dealing with recalcitrant firemen, Hosken's attention was on the whole ship and he spent

Saloon of *Great Western*. (SS Great Britain Trust)

much of his time on deck. The passengers were considerably less of a challenge as there were only seven. Hosken kept an eye on progress across the Atlantic assisted by a special chart, Brunel's attention to detail extending to the charts prepared.

> When we started the *Great Western* to New York, I had a chart drawn and engraved of the sea (that is the lines of latitude and longitude, and the bearings of the compass and the coast and soundings) on a cylindrical projection of the great circle from Bristol to New York; and we found it very useful for the captain to see his great circle sailing and see how much he was deviating from it. [*The Life of Isambard Kingdom Brunel*]

At last after so many days at sea the passengers saw land and they then headed into port. Their first news after so long at sea came from the pilot out of New York who gave them the less welcome information that they were not the first steamship into New York; they had been beaten by the *Sirius*, which was almost half their size. This ship had been hired by a rival company. The competition to be the first across the Atlantic had become intense and one company led by John Junius Smith did have a ship on the stocks, but various problems had occurred to delay the launch. So, they decided to hire a new vessel which was destined for the crossing of the Irish Sea. The *Sirius* set out a few days before the *Great Western* and from Cork, Ireland, which meant a shorter distance. She arrived just a few hours before Brunel's new vessel. New York was

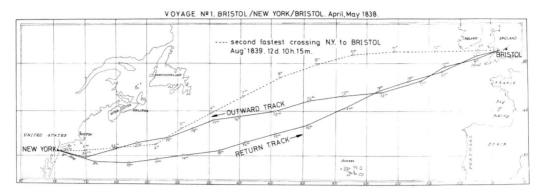

Route of Voyage No. 1. (Denis Griffiths)

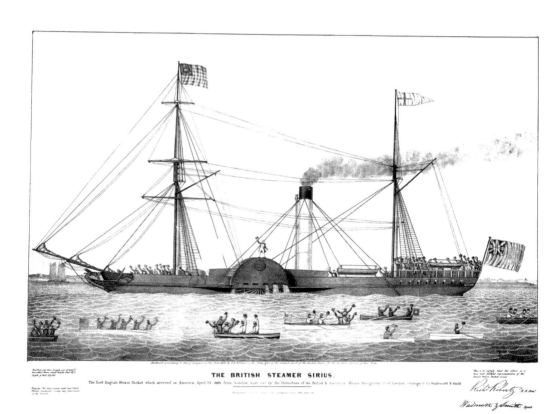

Sirius. (New York Public Library)

absolutely delighted to receive not just one steamship across the Atlantic but two. The crowds were enormous, flags were flying, guns were being fired and an atmosphere of celebration prevailed. It was a great moment in establishing a regular fast link across the Atlantic as Brunel's ship had proved beyond doubt that a steamship could cross the Atlantic in fourteen days with coal to spare. No wonder the Mayor and Corporation of

Crowds greeting the arrival of *Great Western* in New York. (New York Public Library)

Great Western arriving in New York, April 1838. (New York Public Library)

New York were so excited as communication times were halved and information could flow freely between America and Britain. Where the packet ships might at their fastest with the prevailing wind take two or three weeks, they could also take the equivalent number of months, and so were unreliable for fast dispatches of information and people. The two captains of these two steamships, Captain Richard Roberts of the *Sirius* and

Captain James Hosken of the *Great Western*, were feted as celebrities and dined out in New York by local politicians, businessmen and noted residents.

For Brunel this was a very personal triumph, establishing him as an engineer of note, and was even sweeter when it had shown so many self-styled experts to be wrong. It was his first great achievement as the Great Western Railway was still under construction. The *Great Western* became a regular visitor to New York and proved both financially and in service to do exactly what she was designed to do. The ship was famous across the two continents. Songs and music were composed in honour of the ship and dedicated to the directors of the Great Western Steamship Company and Lieutenant James Hosken, the commander.

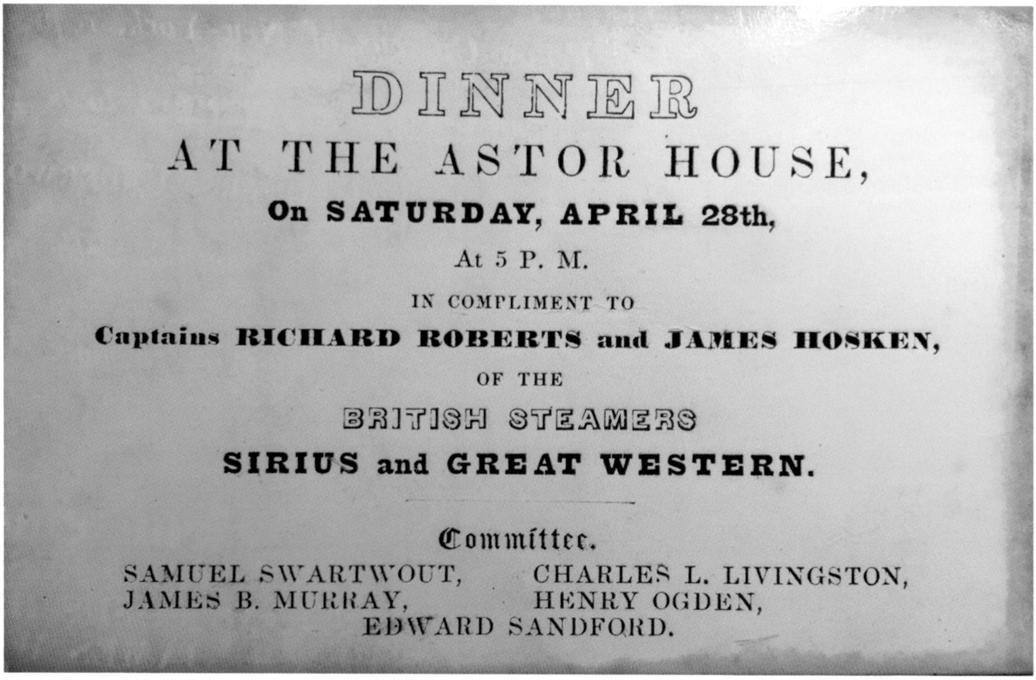

DINNER
AT THE ASTOR HOUSE,
On SATURDAY, APRIL 28th,

At 5 P. M.

IN COMPLIMENT TO

Captains RICHARD ROBERTS and JAMES HOSKEN,

OF THE

BRITISH STEAMERS

SIRIUS and GREAT WESTERN.

Committee.

SAMUEL SWARTWOUT,   CHARLES L. LIVINGSTON,
JAMES B. MURRAY,   HENRY OGDEN,
                 EDWARD SANDFORD.

Astor House Invitation. (New York Historical Society)

Passenger Requirements

Lessons had to be learnt about managing this new type of vessel and these lessons were not just in the engineering department. Managing this new breed of Atlantic liner threw up new challenges in the hospitality department, management of passengers and merchandise. The ship only carried one class of passenger, superior class, at a cost of 35 guineas each. It had not been intended to carry large quantities of cargo, but they had not reckoned with the hand luggage issue. Passengers arrived with large quantities of luggage and eager merchants seeking opportunities for trade brought samples on board with them. All of this arriving as everyone boarded the ship inevitably made for difficulty since it all had to be stowed before departure. The directors of the company announced they would limit what was taken on board to 'portmanteaus or carpet

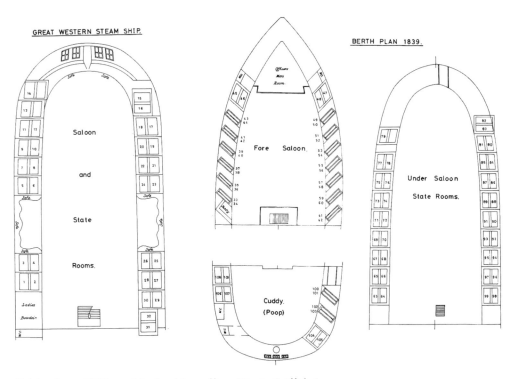

Cabin plans, 1839, used by booking office. (Denis Griffiths)

bags for use on the voyage' and an additional 15 feet of luggage was allowed for each adult while samples of merchandise were charged at a different rate. The hospitality department had to be expanded to meet passenger demands and an early decision to provide free alcohol on board was reversed.

The ship became the equivalent of the Concorde aircraft in its day, attracting the rich and the famous as the fastest way to cross the Atlantic. Wealthy New York families saw the benefits of speed on board the ship and its comparative luxury compared to the sailing packets. James Roosevelt, whose younger son Franklin Delano Roosevelt was later to became President of the United States, was a regular on board with his wife and eldest son. Mr and Mrs Henry James were also great travellers; they were the parents of Henry James the novelist. One interesting character was Prince Murat. His full title was Lucien Charles Joseph Napoleon, Prince Français, Prince of Naples, second Prince of Pontecorvo, 3rd Prince Murat. Lucien's father was a Napoleonic general who had been granted the titles and these went with the fall of Napoleon. Lucien took up residence in the United States but in his frequent European visits tried unavailingly to get his titles and throne restored. One New York couple, Mr and Mrs Dixon, were on their honeymoon trip to Europe in 1840 and Elizabeth Dixon left a detailed account of her voyage in her diary with amusing, and occasionally caustic, comments on her fellow travellers, who included bankers, merchants and army officers. Among the many American businessmen to travel on the *Great Western* was Lewis Tappan. He and his brother established the first viable credit reporting service, which today is Dun & Bradstreet.

Lydia Sigourney was known as the poetess of Harford and she was also an editor and correspondent for *Godey's Lady's Book*, a popular illustrated magazine. Lydia travelled over on a sailing packet to Europe to report on the wedding of the young Queen Victoria

Great Western pottery butter dish with lid. (Clive Richards Collection)

and, on her return, she boarded the *Great Western* with some trepidation. She was not alone as steamships still had much to prove to the cautious traveller. But her fears turned to pride in the steamship and its crew when the vessel was caught in a very large icefield some 50 miles wide. The singular advantage of the steamship was it could reverse its paddles and, eventually, with quite some difficulty, it made its way back out of the icefield overnight. At dawn they could see the vast expanse of the ice field which had entrapped several ships. Lydia wrote a suitably dramatic poem about the incident which was published in the *Lady's Book*. A couple of selected verses will show the type of poetry for which Lydia was noted:

The King of the Icebergs
by Lydia Sigourney

Serene, the Sabbath evening fell
Upon the Northern deep,
And lonely there, a noble bark,
Across the waves did sweep;
She rode them like a living thing,
That heeds nor blast nor storm,
When lo! The King of Icebergs rose
A strange and awful form

Yet on the gallant steamship went,
Her heart of flame beat high,
And the stream of her fervent breath flow'd out
In volumes o'er the sky;
So the Ice King seized his deadly lance
To pierce the stranger foe,
And down to his deed of vengeance rushed
Troubling the depths below.

Not just poets and writers but stars of the stage were keen to come across on the *Great Western* and the cash-strapped New York theatres welcomed them with open arms. The stars of the day included Fanny Elssler, an Austrian dancer; Madame Celeste, an actress; and Madame Vestris. Vestris was not just a successful leading lady, but a shrewd actor/manager in London.

Speed of communication was a major benefit for the government in London. The situation in Canada was troubling at the time with the possibility of a serious border dispute with the United States. Troops were even lined up along the border. Captain Hosken was very proud that his ship was tasked with carrying urgent dispatches and he went in person to deliver them the relevant government minister. Diplomats and civil servants of all nations crisscrossed the Atlantic via the *Great Western*. Speed of news was also of great value to businesses, who could get faster information on the London stock market, cotton prices and much else. Shops in New York could stock fabrics and images of the latest London and Paris fashions; the youthful Queen Victoria was a particular fashion icon of the day. For those who had travelled a distance from home it

Left: Madame Vestris. (New York Public Library)

Below left: Fanny Elssler in Park Theatre dressing room, 1845. (New York Public Library)

Below right: Madame Celeste. (New York Public Library)

THE **MAIN** QUESTION.

Cartoon of Queen Victoria, Lord Melbourne and President Van Buren during the 1838 border dispute. (New York Public Library)

was also a wonderfully speedy way for emigrants to send and receive family news. On at least two occasions the ship was used as a getaway vehicle. Samuel Swarthwout was the Collector of Customs in New York and travelled on the *Great Western* in August 1838, mainly to avoid difficult questions about $1 million which had been diverted into his private ventures. Going the other way a few years later was James Flinn, the newly appointed treasurer to the City of Dublin, who also made a quick getaway with £5,000 of city funds. He travelled to Liverpool to join his wife and caught the *Great Western* across the Atlantic.

The *Great Western* ruled the Atlantic for two seasons before steamship competitors were able to establish lines, but few of the new lines lasted very long. The most serious competitor was Cunard, who won the prize of a government mail contract not to New York but to Halifax. Such mail contracts provided an invaluable subsidy to maintain a regular service in all weathers and all seasons across the Atlantic and the failure to win the mail contract was a major blow to the Great Western Steamship Company. Mail subsidies helped to keep ships on routes for longer seasons and compensated for seasonal variations in passenger numbers. While a non-subsidised company might decline to send the ship across with just a few passengers due to the expense, a ship which was under the mail contract had to go regardless of weather conditions or passenger bookings. That is not to say that the *Great Western* did not carry letters as they regularly carried thousands of letters and newspapers connecting the old and the new worlds.

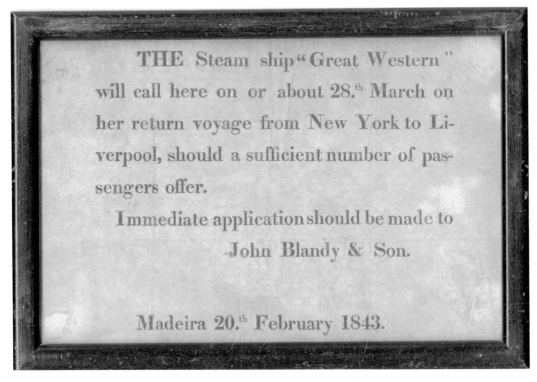

THE Steam ship "Great Western" will call here on or about 28.th March on her return voyage from New York to Liverpool, should a sufficient number of passengers offer.

Immediate application should be made to John Blandy & Son.

Madeira 20.th February 1843.

Advertisement for a voyage from Madeira, 1843. (SS Great Britain Trust)

Great Western in severe Atlantic storm, 1846. (*Illustrated London News*)

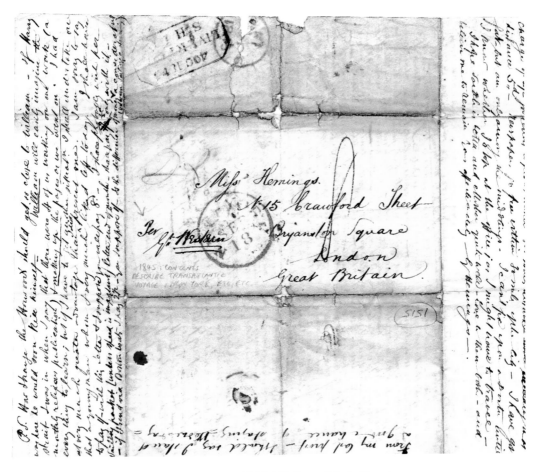

Letter carried on board *Great Western*, 1845. (SS Great Britain Trust)

A Sister Ship

Once the *Great Western* had safely proved its success across the Atlantic, Brunel and the building committee turned their attention to the sister ship. It had always been the intention to have two vessels on the line which could provide a regular service and provide backup. The initial plan had been a simple one: to build an identical ship to the *Great Western*. If they had remained with this plan and spent their money on that sister ship the future of the Great Western Steamship Company might well have been very different. Their reliability would have improved, they would have had a steady income and might well have had a better chance of securing the mail contract. But they did not seem to be in the business of commercialising the route and the earnings of the *Great Western* as she ploughed steadfastly across the Atlantic and the monies raised through the shares of the company went on experimental grand designs as Brunel and his colleagues had other ideas. They saw themselves as pioneers in Atlantic steamship technology, leading the way. From the very first crossing Brunel had been collecting data about the performance of the ship. He regularly put his assistants on board the *Great*

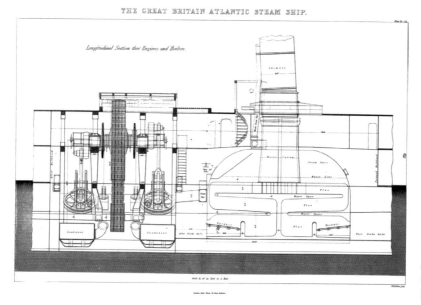

THE GREAT BRITAIN ATLANTIC STEAM SHIP.

1843 engine and boilers. (SS Great Britain Trust)

Western to monitor performance, take detailed notes of how the engine performed and how the ship itself performed at sea. He took a very scientific approach to collecting data. There are two aspects to this – the engines and the ship design – and it was with the ship design that Brunel made the biggest change, based on emerging technology.

Timber was the accepted material for all shipbuilding and a large amount of timber was bought by the company to begin the building of the next paddle steamer. Indeed, they bought sufficient timber to build not just one further vessel but two. But despite this purchase, within months there was change. At the annual general meeting in March 1839 the directors announced that the next vessel was to be built of iron. This was a dramatic change as the use of iron as a shipbuilding material was very much in its infancy.

Iron had been used for boats initially on canals; the *Vulcan* was an iron passage boat which was launched in May 1819 on the Monkland Canal in Scotland. In October 1838 an early iron-built coastal steamer, *Rainbow*, visited Bristol. The building committee of the Great Western Steamship Company was able to examine the ship and to evaluate the practicality and benefits of iron ship construction. One of the benefits was that an iron-built ship of the same dimensions as a wooden ship could carry more cargo. The thinner shell plating and the strength that came from the iron beams and frames could increase the internal space by up to 20 per cent. A considerable benefit was the ability to build ships at greater length. The *Great Western* was the largest ship of its day when it was launched but it had to be very thoroughly strengthened and structured due to the inherent weakness of a long ship made of timber. Brunel noted the comparative benefits of iron in icy sea conditions. He listed other qualities such as the lack of dry rot and he rather optimistically believed, based on what can only have been limited evidence, that there would be no vermin. Having persuaded the directors to change to iron, the major challenge now was the lack of shipbuilding bases building iron vessels. Highly skilled shipwrights were well accustomed to wood, but iron needed to be cut and drilled by hand to a very high degree of precision, and then riveting the plates also required new skills. By the end of 1838 the company decided they would build the ship themselves in their newly acquired dock.

By January 1839, the stated dimensions of the new steamship would be a length of 260 feet, a breadth of 41 feet and a depth of 24 feet. Building began and all seemed set for completion in 1841. During its construction the ship attracted considerable interest with Captain Claxton remarking that 'naval officers, ship builders, engineers and philosophers from all countries began to seek admittance and many have been the papers which in most languages had been written on the comparative merits of iron and wood as material for shipbuilding'. Tenders were sought for the engines and, as before, Maudslay, Son & Field were approached, as were other firms, including a Mr Humphreys. There were many debates over different types of engines to drive the paddle wheels for the new iron ship, but the directors of the company decided, against advice from Brunel, to hire Francis Humphreys and have his design of engine built by the company in their own workshop. Brunel, as the company's consulting engineer, was never happy with Humphreys and their relationship was fractious.

In May 1840 another ship visiting Bristol was to have a dramatic impact on the embryo *Great Britain*. The screw steamer *Archimedes* promised a wholly different method of propulsion. While paddlewheels could provide stability in many circumstances, there were limitations on the high seas. One factor was that as the coal stocks diminished the ship rose higher out of the water and the paddle wheels had greater difficulty in driving the ship forward; similarly, heavy seas might cause one wheel to come out of the water. The propeller did not have that problem. There were two main promoters of the propeller: John Ericsson and Francis Pettit Smith. The Ship Propeller Company was formed in 1839 to exploit Smith's patented location for the screw propeller. It was well funded and included John and George Rennie, who were engineering contractors to the Admiralty. They built the 200-ton *Archimedes* to prove the propeller concept. The ship was sent on a trip around the coast of Britain after her initial trials to publicise the screw and arrived in Bristol in May 1840.

Claxton, the shipbuilder Patterson and Brunel all visited the *Archimedes* and made experiments with the agreement of the owners. Brunel was particularly interested in how to transmit the power from the engine to the propeller. Brunel wrote to Claxton:

> A short time back Barnes, who was with us on the *Archimedes* told Phipps that he had considered the results of our experiments had made the screw better than the common paddlewheel… But that taking into consideration all the advantages of the screw, it was better than any paddle and that he had no doubt it would soon supersede the paddle. I never heard this till today. Phipps is positive of the whole. This is satisfactory. But except to Guppy or Bright [GWSSC directors] do not mention it.

As a result of all the testing Brunel wrote a lengthy, and now famous, report to the directors of the Great Western Steamship Company on the advantages of the screw propeller over paddle wheels. Demonstrating his considerable theoretical knowledge of hydrodynamics, the results of the many experiments demonstrated by the *Archimedes* and the data from the *Great Western*'s performance on the Atlantic, Brunel's report recommended that the new steamer should be adapted for propeller. So, in December 1840, came the second major change to the new ship, as the directors agreed to adopt screw propulsion and Francis Humphreys resigned from the company. Humphreys' engines were no longer required, and Brunel turned to his father's early design for a marine engine. It is then that Brunel became a major figure in influencing the Royal Navy in its decision to use the propeller.

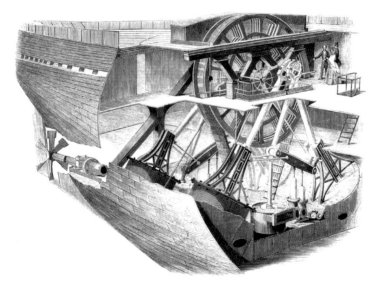

Drawing of the drive shaft and propeller. (SS Great Britain Trust)

Propellers and the Royal Navy

The Brunel family were well acquainted with the navy. Isambard was born in Portsmouth while his father, Marc, was developing the blockmaking machinery for the Royal Navy. Marc continued to act as an occasional advisor to the Admiralty, including the early steam ship trials already mentioned. Until now, Brunel's connections had been indirect, but he was in a different situation now with the success of the *Great Western* and of his London to Bristol railway. He had real status as an engineer.

It has been suggested that the Admiralty was reluctant to move with the times when it came to steam, especially propeller-driven ships. The Admiralty was an early adopter of steam using paddlewheel tugs and packet boats. Steamships with paddlewheels were also used for amphibious assault and inshore bombardment, but they were vulnerable to gunfire in a front-line engagement. The screw propeller held out more possibilities for the navy, but it was not a straightforward decision; there were, as Professor Lambert has pointed out, financial, technical, political, tactical and strategic questions that had to be addressed. The Admiralty was constantly bombarded with suggestions from well-meaning and self-interested parties anxious to promote their invention or idea. The Admiralty's strategy was to keep a watchful eye on developments and to get the private sector to spend their money and time rather than rush into costly experiments.

Administratively there were difficulties as responsibility for these new developments was split between two departments. Captain William Symons was the surveyor of the navy, a dockyard-trained naval architect, but Captain Sir William Edward Parry was the controller of the steam department. Symons largely ignored Parry and there was no integrated approach to steam warship design. This lack of cohesion was only resolved in March 1850 when the steam department was added to the surveyor's office.

To be effective, the success of a propeller rested both on its design and its position within the ship. Promoting different ideas to the navy were Eriksson, Frances Pettit Smith and others who were eager to make money out of their inventions and get the seal of approval from the Admiralty. Meanwhile the navy was taking a keen interest in Brunel's activity; Admiral Parry requested that a copy of Brunel's company report should go to the First Lord of the Admiralty. By bringing Brunel into the process they secured the best engineering advice from an individual who was independent of the competing designs.

Brunel became the leading independent source of advice on the new system through his pioneering work in adapting the *Great Britain* for the screw propeller.

During 1841 the Great Western Steamship Company continued its experimental work with the *Archimedes* led by Thomas Guppy. At the Admiralty, amidst departmental clashes, it was agreed that a new ship would be built due to the importance of the positioning of the screw propeller rather than try and convert an existing naval ship. The navy ordered a new wooden vessel similar to the *Polyphemus*, an earlier wooden ship used for propeller trials, to be built at Sheerness. At Brunel's suggestion the Admiralty used Maudslay for all the machinery. The building of *Rattler*, as she was to be named, was closely supervised by Brunel and Petit Smith and was launched in April 1843 as the world's first screw propeller warship. The ship was towed to East India Dock where Maudslay installed the machinery. The first steam was raised in October 1843 and at each stage Brunel ensured the measurement of trials and engine indicators. Well trained by his father, he wanted consistent measurable data. *Rattler* was involved in thirty-two trials under Brunel's direction for a year between October 1843 and October 1844.

Brunel thus had two propeller trials running concurrently: low-powered naval auxiliary steamers and a high-powered Atlantic liner. These trials led to a reduction in the length of the screw, and longer blades. While the navy adopted a two-bladed hoisting screw, Brunel designed a six-bladed fixed propeller for the *Great Britain*. Throughout this time of experiments Brunel collaborated closely with Francis Pettit Smith, with

Rattler's propeller, in the Being Brunel Museum. (Science Museum Group/SS Great Britain Trust)

whom he developed a good working relationship. Brunel's contribution to the successful adoption of screw-propelled warships was significant and in return he gained considerable information from the many naval trials of the propeller, information that he put to good use. It also confirmed him as 'the leading professional engineer of the age'.

Replica propeller on the *Great Britain*. (SS Great Britain Trust)

Launch of the *Great Britain*

All these naval trials were of immense value to the final building of the *Great Britain* and the design of its propeller. At last on 19 July 1843 she was ready to come out of her dock. It was a grand occasion and Prince Albert himself came down on the new Great Western Railway to view this new ship and to give it his blessing. He was invited to perform the official launch, but he gallantly deferred to Mrs Miles, the wife of one of the directors of the company, who had launched the *Great Western* just six years before. Unfortunately, she missed hitting the ship and Prince Albert, to great applause, picked up a bottle of champagne and scored a direct hit.

Once floated out in to the harbour, the ship, now 40 feet longer than originally proposed, remained there for the next year as she was fitted out. The final design of the propeller had not yet been agreed as they were still waiting for the experiments to be completed on the

Crowds at the launch in the presence of Prince Albert, 1843. (SS Great Britain Trust)

The Launching of the "Great Britain"
19th of July, 1843

The *Great Britain* launched into the floating dock. (SS Great Britain Trust)

Rattler and in the event the propeller had to be designed and fitted without that final data. The big problem now was getting the ship out of the harbour and up to London. All along the company had been optimistic that the dock company would be prepared to widen the walls of the lock from the floating harbour into the river, but this was a major task and an expensive one. Optimism was not met with success as the dock company stubbornly refused to accept any liability and suggested it would need to go through an Act of Parliament. For six months the arguments went backwards and forwards until an agreement was reached. At last on 10 December the ship was ready to move but it was almost a total disaster. The widening was not sufficient and the ship became wedged in the lock – a potential calamity. Fortunately, they were just able to bring her back and subsequently widen the entrance again so that on 12 December, at long last, she was towed down the Avon River and as she went the boilers were filled, the fires lit, and steam was raised.

An anonymous Bristol poet wrote in celebration of the occasion

Six masts – like princely sons to bear!
Great Britain for my name
My smoke trail black on the sun bright air
My screw as swift, and my sails as fair
As the trumpet voice of fame!

Great Britain at last on its way down the Avon. (Joseph Walter/SS Great Britain Trust)

Several trials were carried out locally to test the engines and propeller including a run down the Bristol Channel to Ilfracombe and back. Captain Hosken was in charge having transferred from the *Great Western*. As Ewan Corlett wrote: 'indeed *Great Britain*'s performance on trials was remarkable. There was little or no difficulty with her machinery and no need for alterations, the ship met her intended speed, she steered well; and everything was in order in spite of the novelty of the construction of her means and of her means of propulsion.' He went on to say, 'only naval architects and ship builders can appreciate the enormous achievement this represented in the technological context of the day.'

After these local trials, on 23 January 1845 the *Great Britain* sailed for London, causing a considerable sensation as she passed other vessels on the Thames. The decorators had done their work well with columns of white and gold, pilasters in the saloon painted with oriental birds and flowers and carved and gilded archways to some doors. All the saloons were covered in Brussels carpet purpose-made by Messrs Mogg of Bristol. While the ship was at Blackwall she became, as her predecessor had been, a great attraction, with thousands rushing to see this amazing sight. This time she was visited by the Queen and Prince Albert. The *Great Britain* remained in the Thames for a lengthy five months before at last setting out from Liverpool on 26 July 1845 on her maiden voyage across the Atlantic.

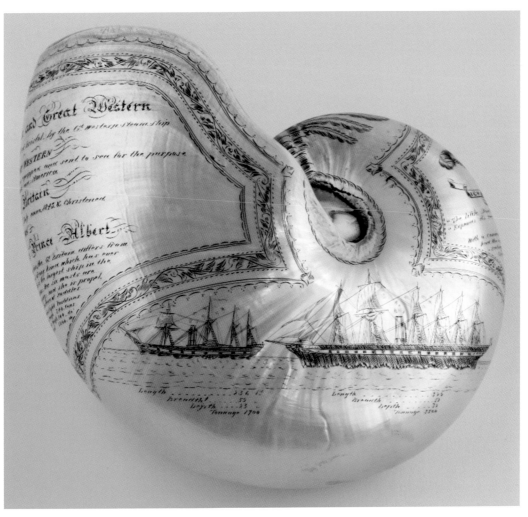

Nautilus shell presented to Queen Victoria and Prince Albert showing the *Great Western* and *Great Britain*. (SS Great Britain Trust)

The *Great Britain* in New York

New York had been expecting the new ship for some time and perhaps the delays in the whole project led to the following rather caustic remark in the *New York Evening Post* on 2 June 1845:

> The *Great Britain* steamship will leave Liverpool on 25 July for New York. Those who expect to see a beauty of a boat, or much fine carving and gilding, will be disappointed. She is built for use, not show, and her only claims on attention are, her stupendous proportions, and her qualities as a sea boat.

The number of passengers on this maiden voyage was disappointing, just forty-five people, but overall it was an uneventful passage. Just before their arrival in New York the passengers organised a meeting onboard on 7 August. They formed a committee chaired by some distinguished passengers. The chairman, Lieutenant Colonel Everett, was a Fellow of the Royal Society, an organisation of which Brunel had long been a member. Other committee members included two US Navy officers, S. K. Kane and Lieutenant H. A. Wise, plus Captain Morris of the Royal Navy and a retired British Army officer, Septimus Crookes. Conscious of history, and no doubt of good publicity, a letter was composed to be signed by all passengers on board. It congratulated the Captain and the company 'upon the successful result of this, the first practical attempt to cross the Atlantic in a vessel propelled by the Archimedean screw propeller.' They praised the size of the ship and...

> ...the nature of her materials, which, taken in conjunction with the character and machinery and the novelty of its application, gave rise to an excited state of public opinion, which attached the highest experimental importance to the successful termination of our passage.... we feel especially called upon to allude to the fact, as interesting to the admirer of the vessel, as important to our own comfort, under the influence of an ordinary breeze, there is toward the head of the vessel absolutely no vibration whatever caused by the machinery; that the vibration at the engine and inward the central part is reduced to a mere tremulous motion; and that even toward the stern where the greatest effect might be expected, it is far less than is usually experienced in vessels propelled by paddle wheels.

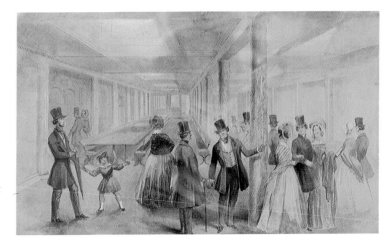

Passengers in the first
class dining saloon.
(SS Great Britain
Trust)

They finished by expressing their 'entire satisfaction with the luxuriant supply of the table and the excellent arrangement of the steward's department'.

This was duly reported in the New York papers whose attitude had changed considerably when the ship arrived. 'Arrival of the Monarch of the Ocean, She is truly beautiful,' exclaimed the newspapers and thousands crowded the shores to catch a glimpse. Just like her predecessor the ship was a major attraction, visited by thousands who flocked to see this new wonder. The *New York Daily Tribune* commented: 'We admired the plain and solid style adopted in all parts of the *Great Britain*, so simple, so judicious, so easily kept clean, so truly English!'

Captain Hosken. (SS Great Britain Trust)

The ship remained in New York for nineteen days and an estimated 21,000 people visited the ship. A dinner was given on the last evening for Captain Hosken. It was hosted by the New York establishment including the Chief Justice Samuel Jones; diarist Philip Hone; James De Peyster Ogden, merchant, President of Brooklyn harbour, and the Nautilus Insurance Company; William Starr Miller, Congressman for New York; and F. H. Delano, a wealthy New York merchant. The ship was the toast of New York and the dinner guests in typical style filled their glasses to drink the following toasts:

The merchants of Bristol – the first to risk their wealth to transatlantic steam navigation. The thanks of both nations are justly their due.

The President of the United States.

The Queen of Great Britain and Ireland.

Captain Hosken, whose skill and deportment have secured the confidence of the public with a well-deserved popularity. May the measure of success correspond with the magnitude of his command.

The memories of Watt and Fulton – in the *Great Britain* we witness the grandest triumph of their art, and the proudest moment of their genius.

The Pacific influence of steam – it makes all nations neighbours, and neighbours should never quarrel.

The cities of Liverpool and New York - Honourable competitors in commercial enterprise.

The ship returned to Liverpool with fifty-three passengers, 1,200 bales of cotton and other merchandise. There were difficulties on this return passage with the engines not giving full power, the loss of a main topmast which snapped, and although the lack of vibration was praised by the passengers, the same could not be said for the tendency of the ship to roll from side to side.

On 27 September the ship left Liverpool and set out across once again to New York. There had been hints previously given of problems with the compass, which 'do not act as perfectly as could be desired'. And it was on this passage that the ship encountered her first navigational problem when she touched a shoal off Nantucket. She was eventually extricated from this unfortunate position. When the ship was dry-docked at New York, an expensive operation, it was found that the propeller had been extensively damaged. After repair the ship returned only for the propeller to break again so that the voyage was completed under canvas.

During the winter, alterations were made to pumps and valves to provide better air flow, which provided more steam, and the ship was fitted with a new screw propeller. Her second season commenced on 9 May but with a small complement of passengers, just twenty-eight; this increased on 8 June to forty-two and the best numbers were on her passage from Liverpool in July when she carried 110 passengers. On this passage Hosken again had a navigation problem when in thick fog the ship touched a reef off Newfoundland. Despite this the ship was showing real promise that she could make regular passages of thirteen days out and eleven days back. This was not to be realised.

Disaster

On the morning of 22 September 1846 in moderate weather the ship left Liverpool, having on board 180 passengers – her largest passenger set yet – and a considerable quantity of freight, all promising future financial benefits for the cash-strapped company. She passed south of the Isle of Man and would, as normal, plan to head north around Ireland; instead, in an inexplicable failure of navigation, she ran ashore on a rocky beach at Dundrum Bay in Ireland at 10 o'clock that night. The passengers panicked and wept, prayers were said but nothing could be done until daylight. At dawn all passengers were safely lowered over the side of the ship and conveyed in carts to find shelter.

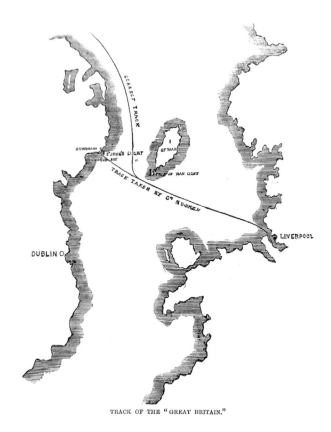

The fateful track to Dundrum Bay. (*Illustrated London News*)

TRACK OF THE "GREAT BRITAIN,"

It was a total, very public and embarrassing disaster and a public relations catastrophe. While the company directors loyally blamed navigation charts, other commentators put the blame on Captain Hosken. The *Mechanics' Magazine* was firmly of the view that 'The case was on his own showing and beyond all possibility of doubt the most egregious blundering.'

Claxton rushed to see the ship for himself. She was in one piece but had settled onto some rocks which had knocked holes into her. The following Monday in the spring tides Claxton tried to get the ship off but a gale of wind prevented the operation and the decision was taken to drive the ship higher up the beach where she would be slightly less exposed. 'Sails were therefore set and she was driven forward a considerable distance.' The shipbuilder Patterson was now sent by the directors to Dundrum and timber breakwaters were built to try to protect the ship, but these were soon carried away by gales.

The Great Western Steamship Company had been experiencing financial difficulties for some time, largely due to the ever-increasing cost of building the *Great Britain*. They had pinned their hopes on securing subsidy through a government mail contract, to no avail. The stranding of the *Great Britain* sealed the fate of the *Great Western* steamship. Due to the rising cost of the building of the *Great Britain*, it had been on the market for a while but now the company urgently needed to raise funds to salvage their new flagship. In 1847 the *Great Western* was sold to the Royal Mail Steam Packet Company and proceeded to provide a successful mail service to the West Indies and South America for many years under her new owners.

In December at last Brunel was free from his Parliamentary work and was able to go to Dundrum to see for himself and to report on the ship. To his very considerable relief he found that the *Great Britain* was in his words 'as sound as the day she was launched in 10 times stronger and sound and character.' He was determined his ship would be rescued and wrote what his son later described as a somewhat vigorous letter to Captain Claxton. The most famous statement in that lengthy letter is his fury that 'the finest ship in the world, in excellent condition, ... has been left lying, like a useless saucepan kicking

Great Britain grounded at Dundrum. (*Illustrated London News*)

about on the most exposed shore that you can imagine, with no more effort or skill applied to protect the property than the said saucepan received on the beach at Brighton.'

Claxton now came under a stream of instructions from Brunel to protect the ship. Instead of a solid timber breakwater, vast bundles of faggots and sandbags were to be used to defend her from winter storms. Claxton complied with Brunel's demands but had initial difficulties in fixing the new defences. Brunel was brutal in his reply to Claxton: 'You have failed... nine tenths of all failures in this world [come] from not doing enough.' Claxton redoubled efforts and quantities and finally succeeded in safeguarding the ship through the winter. The next major question then was how to get the ship safely away. Every possible item was removed to lighten the ship and the salvage experts James Bremner and his son Alexander were brought in to assist. From May 1847 for several months the world looked on as the salvage party worked to rescue the great iron ship. The *Illustrated London News* carried a constant series of drawings with news of progress as the ship was lifted, wedges and stones were rammed underneath to raise her and the main leak was temporarily repaired. At last in August 1847 Claxton could write to Brunel, 'Huzza! Huzza! You know what that means...'

The ship was finally released from her enforced imprisonment on the beach, while ironically promoting the durability and strength of iron as a shipbuilding material. Such was the importance of this vessel that the Admiralty had sent two ships, *Birkenhead* and *Scourge*, to assist, plus men from the dockyards at Portsmouth and Plymouth. With considerable effort the leaking ship arrived in Belfast, where she was temporarily grounded. The next day, with the pumps working the whole time, the ship was taken across to Liverpool. It was a spectacular marine salvage event.

Meanwhile the company, having saved the *Great Britain*, now had to realise that this was the end too for them. The ship was repaired and then put up for sale. After a few failed attempts she was finally sold for the bargain price of £18,000 to Gibbs Bright of Liverpool in December 1850. Robert Bright had been a key supporter and encourager of Brunel's vision for his two ships and had been not just a director of the Great Western Steamship Company, but at one time its chairman. The Gibbs Bright company had also been the ship agents for the *Great Western* and *Great Britain* in Liverpool. Their interest in the ships was for a new opportunity opening in trade with Australia.

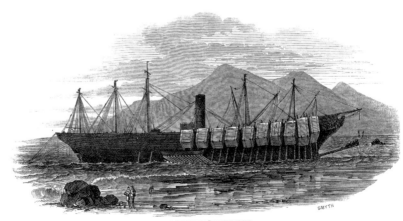

The salvage operation. (*Illustrated London News*)

THE "GREAT BRITAIN" BEGINNING TO RISE.

The Australia Years

The discovery of gold in Australia in Victoria in 1850 had a dramatic impact and there was a rush of emigrants keen to make their fortune in the gold fields. Ships of every description were needed to cope with demand. Gibbs Bright already had good Australia connections, running the Eagle Line of sailing packets, and the first ship to arrive in Liverpool with gold from Australia was the Eagle Line packet *Albatross*, landing £50,000 worth of gold dust. As Corlett says, 'casting around for tonnage, Gibbs Bright who had been the Great Britain's agents on the Atlantic run thought of the ship. Here was a magnificent vessel, the largest in the world, capable of taking considerable numbers of passengers – not to mention cargo – each way and going cheap.'

Gibbs Bright were interested purely in a commercial venture. The Australia run was hard on sailing ships. The ships ran with the trade winds down the Atlantic almost to South America and then turned east towards to the Cape of Good Hope and here in the southerly regions they could encounter high seas, strong headwinds and ice. The *Great Britain* was a tough well-built ship with good sailing qualities; the sails could be used for much of the time and the new engines by John Penn & Company provided power when required.

William Patterson, the Bristol shipbuilder, was put in charge of the conversion. With a long passage to Australia significantly more cargo space was needed. The total number of passengers was now 730, of which fifty were first class. There were ladies' boudoirs for privacy and an innovation was the hurricane deck, a pleasant space in tropical climes. The grand saloon was 75 feet long and the *Illustrated London News* approved of the tasteful decorations. This saloon was also very popular in the colder climes since it was one of the warmest spaces on the ship.

But before she went to Australia it was decided that she would do one further trip across the Atlantic. Hosken's once great reputation was severely injured and by now he had served as master of his last merchant ship. Barnard Matthews, his long-serving second officer on the *Great Western* and latterly master of that ship, was appointed captain with John Gray as his second officer. So, on 1 May 1852 with 180 passengers the *Great Britain* sailed again, but not initially for Australia as she headed for a run to New York, doing well despite strong gales and arriving in thirteen and a half days.

Barnard Mathews, his officers, cadets and Francis Petit Smith, patent holder of the screw propeller, on *Great Britain*. (Clive Richards Collection)

1852: The First Australia Run

On her return she now was ready to make her maiden voyage to Australia, carrying 630 passengers on board, a crew of about 150 and a very considerable amount of mail. It was the largest complement of passengers she had ever carried. She was anticipated to reach the Cape in twenty-five days and then Melbourne in fifty-six days, halving the time of the average sailing vessel. In case of any attack the ship had six deck guns and was well supplied with arms and ammunition.

Captain Matthews, whose experience of ocean travel had been solely across the Atlantic, seems to have been fazed by the navigational challenge of a very different route. The *Great Britain* had the advantages of steaming through the doldrums but met severe head winds, forcing it to turn back when it was discovered that coal supplies were alarmingly low. This involved a significant detour back to St Helena for fuel. There was much criticism of Captain Matthews by his passengers, some of whom were seamen, and instead of the expected run of sixty days she took eighty-three days to reach Melbourne.

The *Great Britain* was greeted with considerable warmth by the Australians and on her arrival in Sydney on 20 November she was greeted by more cheering crowds and the governor and his family made a visit. The Australian newspapers were full of praise. Just as New York had welcomed the arrival of the *Great Western* in 1838, the arrival of the *Great Britain* offered to cut journey times significantly and promised a regular fast connection back to Britain. On her return to Britain it was realised that more alterations were required, and that the vessel was underpowered.

Her second voyage to Australia was in August 1853 and this time she made Melbourne in sixty-five days, carrying thirty-four first-class passengers, 119 second class and 161 third class. On her return to Liverpool she carried 199 passengers and made it back in sixty-two days. She also returned with 7 tons of gold, twenty-three bales of cotton plus more tin and 15 tons of mail. Matthews, who had been heavily criticised by the passengers as indeed he had when master of the *Great Western*, at this stage decided to retire to Australia and his number two, John Gray, took over. But Matthews had proved under his command that the *Great Britain* could succeed in her new role in accomplishing the run to Australia and back in just three months.

Right: Poster advertising passage on the *Great Britain* in 1873. (SS Great Britain Trust)

Below: Painting of the *Great Britain* in Melbourne harbour. (SS Great Britain Trust)

Crimean War Service

In July 1854 the ship was now managed by the Liverpool & Australian Steam Navigation Company, a subsidiary of Gibbs Bright. But before she could go back to Australia she was chartered by the government for war duty. The Crimean War, known at the time as the Russian War, was in severe need of transport ships and the government chartered mail ships. The British and the French fleets had entered the Black Sea in January 1854 and the invasion was planned for that September. Large numbers of ships, both sailing

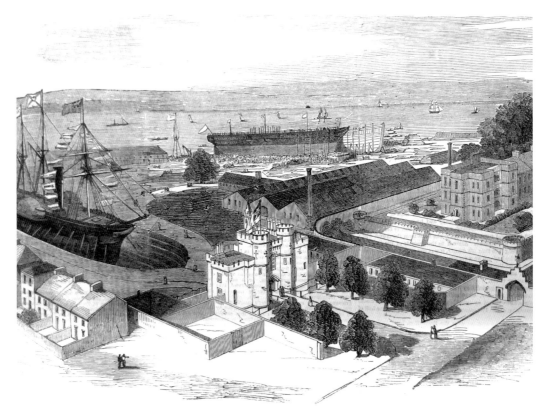

Great Western at Northfleet, 17 May 1851, and *Orinoco* dressed overall ready for launching. (*Illustrated London News*)

NAVVIES RECEIVING AND CARRYING THEIR KITS ON BOARD.

Large numbers of labourers were taken to and from the Crimea. (*Illustrated London News*)

The small harbour at Balaclava. (*Illustrated London News*)

ships and steamships, were under government contract. The *Great Britain* was ideal as a troopship and she carried 1,650 infantry and thirty horses. She continued to move backwards and forwards between the ports from Gibraltar to Balaclava and by the time her service finished in June 1856 she had carried nearly 45,000 troops. She was not the only one of Brunel's ships in the Crimea as the *Great Western*, now owned by the Royal Mail Steam Packet Company, was also under government contract.

To carry troops and horses to the Crimea the ship had been stripped of much of her passenger accommodation and at the end of her war service considerable work would need to be done to bring her back into a fit condition for passengers. In October 1855 she was offered for sale to the Admiralty for the transport service, but they declined. In 1857 the *Great Britain* changed shape yet again with new machinery, new lifting gear for its propeller and changes to the decks. She now had one funnel and three very much heavier masts and looked externally quite different to the ship launched by Brunel. After nine months of work the ship arrived in Melbourne in 1857 with a good passage of sixty-two days, but her war service was not over. The Indian Mutiny broke out in 1857 and the *Great Britain* was rushed out with troops for Bombay together with sixty-eight other troopships. She took 48,482 men of the 17th Regiment of Lancers.

Her next voyage after this was in fact not to Australia but, oddly, it was a last trip to New York. Her final arrival in that great port was in 1858 and she returned to Liverpool on 9 September 1858 with seventy passengers, 60,000 dollars in specie and a full general cargo in twelve days. Following this Atlantic voyage, the *Great Britain* now began a regular run to and from Australia, achieving it in a steady record time.

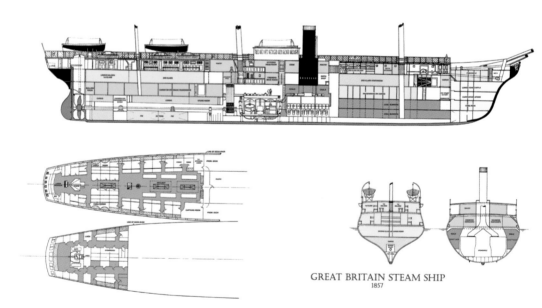

GREAT BRITAIN STEAM SHIP
1857

The layout of *Great Britain* in 1857. (SS Great Britain Trust)

Passengers to Australia

Unlike her sister ship, the *Great Western*, which had but one class of passenger, the *Great Britain* had first, second and third class. There were distinctions between each group in terms of their recreational spaces, both internally and externally, and in their cabins and their food. Rachel Henning travelled on the *Great Britain* in February 1861. Aged thirty-five and single, this was her second voyage to Australia, where she had a brother and a sister. Rachel's previous passage had been in a sailing ship. She wrote home to her family describing the difference.

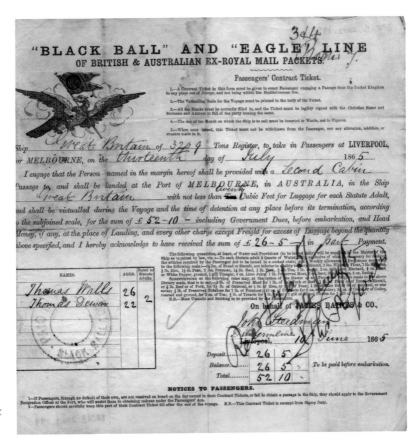

Second class passenger ticket. (SS Great Britain Trust)

"GREAT BRITAIN,"

Live poultry on the deck of the *Great Britain*. (SS Great Britain Trust)

The captain sits at the top of the dining-table, next to the mast. Mrs Bronchordt sits next to him at his right, and next to her I sit; then a Mr Brand, a Scotchman; the pretty German and her husband sit opposite, and the 'commercials' down the same side. I can tell you nothing about the inhabitants of the different cabins, of course, I know none of them. I rather like a stout good-natured woman, who inhabits, with her husband, the one opposite to ours; but she is not a lady. You cannot think how dirty everything gets; hands, clothes everything is black. The white in my dress is in a most disastrous state. I never saw such a dirty ship.

A database has been created by the Brunel Institute at the SS Great Britain Trust of all the passengers and crew who travelled on board the *Great Britain*. Analysis shows that 30 per cent of passengers over the age of sixteen were women and their average age was thirty. Most of them had professions such as seamstress, dressmaker, or governess. Agnes Broadbent, a forty-eight-year-old second class passenger who travelled in 1853, is a little more unusual as her occupation was listed as an ironmonger.

Other passengers included Edmund Veness, age thirty-four, his wife Elizabeth, age thirty, and his sister Sarah, age forty, all of whom travelled out in April 1863. Edmund was a gardener and was at one time listed as a curator working in Queen Victoria's gardens. Edmund's wife, Elizabeth, died one year after their arrival in Victoria but Edmund remained and bought 29 acres of land from the government in 1866, cleared it and established a successful fruit farm. Ten years later he remarried aged forty-eight and had five children.

Watercolour
of the deck of
Great Britain
by a passenger
in 1873.
(SS Great
Britain Trust)

The majority of passengers were male, often single and setting out to make their fortunes in a new country. Edward Towle travelled second class in 1852. He paid between 25 and 40 guineas and he hoped to work in farming or gold prospecting. He commented on his fellow passengers:

> There seems to be a great mixture of characters on board, men who had been gambling the night before now appeared at church with a most devotional demeanour, and others who appeared to be very steady and sedate never went to church at all... We have French, Germans, Poles, Jews, Italians, Scotch and Irish on board.

Allan Gilmour provides a rare glimpse of life as a steerage passenger. At the age of seventeen he travelled with his father and brother, in cramped conditions with four people per cabin, and he also noted the considerable gambling on board with as much as £500 changing hands in one game. But it was not just emigrants who travelled out to Australia as the first All-England cricket team was on board in 1862. Two Melbourne-based businessmen, Felix Spiers and Christopher Pond, had invited them. Twelve players signed up and received £150, first class passage (70 guineas) and all expenses. It sparked an Australian passion for the sport and the England team won six games, drew four and lost two on rough grounds, some of which had been hurriedly created for the occasion. It was the first All-England cricket tour, the first commercially sponsored and the first time the term 'test match' was used.

In 1872 there was a tragic incident on board, which left a mystery that is still unresolved. The ship, under the command of Captain Gray, left Melbourne on 23 October. In the South Atlantic it was discovered that Captain Gray had disappeared. He had joined the ship in 1852 as second officer, served as the master of the ship for eighteen years and had been a very popular and highly effective captain. The mystery was that his servant went to call him and found his cabin empty with the only evidence of his demise being an open window in the stern of the ship. The first officer, Robertson, had to take command. On arrival in Liverpool, he was not confirmed in the position; the second officer, Chapman, now became the fourth master of the *Great Britain*. However, her days as a passenger ship were numbered and, thirty-three years after her launch, the ship returned to Liverpool in February 1876 on the very last passage from Australia.

Left: Captain Gray, long-serving captain of *Great Britain*. (SS Great Britain Trust)

Below: Great Britain in 1872. (SS Great Britain Trust)

A Cargo Ship

For five years the *Great Britain* lay idly moored at Birkenhead until in July 1881 she was put up for auction. At the auction, while attracting considerable attention due to the ship's well-known history, she did not achieve her reserve price as the highest bid was just £6,000. By now Gibbs Bright & Company was Anthony Gibbs, Son & Company, having been absorbed by the London arm. Anthony Gibbs mainly traded with North and South America. Unable to sell the ship for a suitable price, the decision was taken to use the good-sized vessel as a cargo ship. Her interior was completely stripped out, including her engines and funnel, large cargo holds were built, and she became a sailing cargo ship. She was reregistered in Liverpool in November 1882 as an iron sailing ship and her new master was James Morris. Morris did not last long and appears to have

The Historical old "Great Britain" now used as a Wool Store Hulk by the Falkland Islands Company, Ltd., Stanley, Falkland Islands

The *Great Britain* as a wool store in Port Stanley, Falkland Islands. (SS Great Britain Trust)

been demoted to first mate when Captain Henry Stapp became the ship's last master. Her new destiny was as a coal carrier and she loaded 3,290 tons of coal for San Francisco, but she did not get very far as the ship began to leak considerably and had to return.

She eventually sailed for San Francisco in December but had to put into Montevideo due to problems with the stowing of the cargo, which had caused the ship to be unstable. The ship arrived finally at her destination on 2 June 1883 after a 180-day journey from Britain. She was, however, still held in high regard with the local paper referring to her as 'this famous old ship'. She sailed for home one month later, carrying wheat for Cork, and again it was a long journey, this time taking 154 days. She departed in May 1884, again with coal for San Francisco, taking 160 days and weathering severe storms. These passages around Cape Horn were hard on ships and men as they faced the severe weather conditions in the South Atlantic. She returned with wheat for Cork in Ireland in July, making it back in 145 days. Her last voyage as a cargo ship was on 6 February 1886, leaving Cardiff with coal for Panama. She met severe gales which were so bad that the coal shifted and, with intense labour, the crew had to re-level the ship, but no sooner had they done that than another gale struck, which became a full hurricane. The captain agreed with the crew that the ship had to turn back, and they set course for Port Stanley in the Falklands. The ship was written off by the insurers and the vessel was left in the harbour, her sailing days now over.

Brunel's Shipping Projects after 1850

The sale of the *Great Britain* to Gibbs Bright in December 1850 and the end of the Great Western Steamship Company came at a time when Brunel's railway engineering career was changing. In the 1850s, as railway mania had come to an end, there were few new developments and engineers were being laid off. Brunel still had projects on which to work, including his great bridge at Saltash, but this was the period when he began to consider other things, including plans for his new house and estate in Devon. Between 1847 and 1858 Brunel made a series of purchases on land outside Torquay at Watcombe. This was near the coast and perhaps he saw himself as indulging in that rare pursuit for him, leisure. He even sketched a small rowing boat.

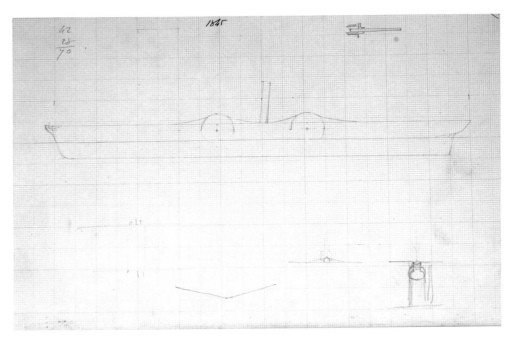

Sketch by Brunel of a steamship, 1845. (Brunel Institute, SS Great Britain Trust)

It might seem that Brunel lost all interest in ships for a while, but this is misleading. After the exertions of her rescue from Dundrum Bay and the subsequent sale of *Great Britain* to Liverpool, Brunel had no further official involvement with that ship but he was kept informed by his loyal friend Captain Claxton. He wrote with pride to Brunel in March 1854 on the condition of the bottom of the ship: 'There is not from stem to stern one single speck of rust.' While this was no doubt of interest to Brunel, he now had other major shipping projects in hand. His sketch books show his continued thinking about ship design and the business world was still eager to use him as a consultant engineer. In the year following the sale of the *Great Britain*, a new company was registered, the Australian Royal Mail Steam Ship Company. The company's chairman was William Hawes, who was the younger brother of Benjamin Hawes, Brunel's close friend and brother-in-law.

Brunel's involvement with this company is usually mentioned as just a simple sentence or throwaway remark but this was to be an important step on the path to his final great ship. Brunel's son, Isambard Junior, explains this in his biography of his father.

> …after the release of the *Great Britain* from Dundrum Bay, Mr Brunel became again connected with the construction of steamships. In that year he was consulted by the directors of the Australian mail company upon the class of vessels which it would be advantageous for them to purchase, in order to carry out their contract for the conveyance of the mail to Australia. He advised them to have ships of from 5,000 to 6,000 tons burden, in order that they might only have to touch the coal at the Cape. Some of the directors would not hear of so startling a proposition; they nevertheless asked Mr Brunel to become an engineer and he retained the post till February 1853. Two ships were built under his direction by Mr J Scott Russell — the *Victoria* and the *Adelaide*.

The Australian Royal Mail Steamship Company lists Isambard Brunel among its most senior staff as their consulting engineer. In a very short time it won the government mail contract for Australia, perhaps assisted by the fact that Benjamin Hawes was the Under Secretary for War and the Colonies, although he resigned from this position in October 1851. The steamship company now needed to fulfil its obligations under the mail contract and initially bought its first vessel. The steamship *Melbourne* was built by Scotts of Greenock in 1849 and was acquired by the company from the Admiralty. It was quickly adapted and sent off, but it very rapidly ran into problems and the Admiralty agent on board had to force the ship into Lisbon for significant repairs.

The *Adelaide* and *Victoria* were commissioned from the shipbuilder Scott Russell by Brunel and were built in Scott Russell's Millwall yard. These were identical ships and they set off at different dates for Australia. The *Adelaide* experienced several difficulties. She sailed from Plymouth on 3 January and arrived at St Vincent on 17 February, then stopped again at St Helena. She reached the Cape in March and stayed there four days. She only used steam for a third of the time after leaving the Cape and some of the second-class passengers were very dissatisfied with their long voyage. The chief engineer was sacked and the third engineer promoted in his place. The second ship, the *Victoria*, went out one year later and was deemed to be a highly successful vessel, but it was too late as by then the government had cancelled the company's mail contract.

In Brunel's sketch book there is a sketch of a ship for the Australian Steam Navigation Company (ARMSC). Brunel drew several sketches comparing outlines: *Wave Queen*, *Great Britain*, *Victoria* and another unnamed vessel. These are of the same dimension, but also in his sketchbooks increasingly there are drawings of much larger ships.

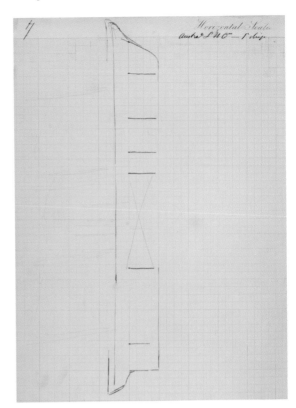

Right: Profile sketch by Brunel of
Australian Steam Navigation Company
ship. (Brunel Institute)

Below: Detail of ARMSNC ship.
(Brunel Institute)

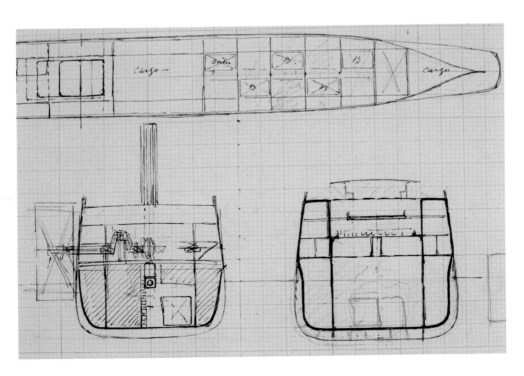

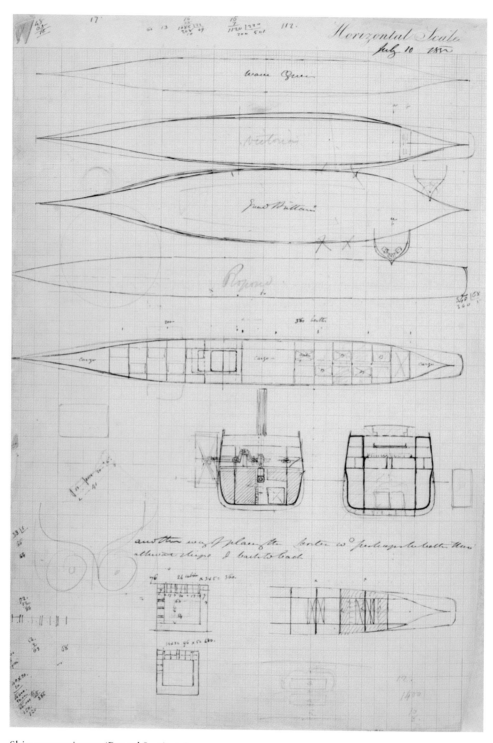

Ship comparisons. (Brunel Institute)

The Great Ship

The *Great Eastern*, described by another Brunel biographer, Professor Angus Buchanan, as the most contentious feature of Brunel's career, has since caused major debates between historians, who are split in their views on whether the various calamities that befell the ship are due to John Scott Russell or Brunel.

While he was consulting engineer for the ARMSC, Brunel considered the challenges of the Australia run. The P&O Company favoured steam through the Mediterranean, then carriage overland to Suez, across the Indian Ocean and then on down to Australia, and they had the monopoly on the mail contracts to India. The ARMSC was travelling via the Cape of Good Hope and their ships needed to stop frequently for coal. Coal had to be supplied expensively from South Wales to the coaling points and was not easily available in Australia for the return voyage. Brunel wanted to make the long steam voyages both speedy and economical. He believed that it required a 'Vessel to be large enough to carry the coal for the entire voyage at least outwards and ...then for the return voyage also.'

If the ARMSC was averse to the idea of giant ships, there was another company who was prepared to listen. The Eastern Steam Navigation Company (ESNC) was formed to win the contract for mail to India, China and Australia but lost out when the government awarded it to the safe pair of hands of the P&O. Having established itself and raised some capital the ESNC was without a purpose, but on 1 December 1852 the chairman of ESNC, Henry Thomas Hope, announced the building of a new large ship.

Brunel had collected information related to trade with India and Australia which demonstrated the advantages to be gained by a rapid and direct communication for the carriage of passengers and troops, as well as merchandise. His plan was for one large ship to carry its own coal and be so large that mail subsidies were unnecessary. The ship also needed to be relatively shallow as his plans included the navigation of the River Hooghly to Calcutta. Much of the following information comes from Isambard Brunel Junior's biography of his father. The family was to remain sensitive to Brunel's reputation in relation to the great ship.

> In February and March 1852 I matured my ideas of the large ship with nearly all my present details, and in March I made my first sketch of one with paddles and screw. The size I then proposed 600' x 70' and in June and July I determined on the mode construction now adopted of cellular bottom; intending to make the outer skin of wood for the sake of coppering.

The ESNC set up a committee to interview both Brunel and Scott Russell. Scott Russell, 'who was fully acquainted with all Mr Brunel's plans and had ably assisted him in

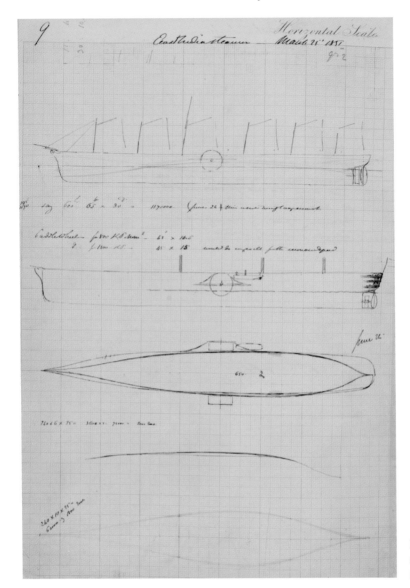

East India
steamship.
(Brunel Institute)

maturing them,' attended and spoke on behalf of Brunel and Brunel was duly appointed engineer to the company. In July 1852 Brunel wrote to the directors and recommended that two identical ships should be made and that 'they must in fact be exact duplicates of each other'. He then proceeded to list the men with whom he was going to work.

...with respect to the construction of the vessel itself, nobody can, in my opinion, bring more scientific and practical knowledge to bear than Mr Scott Russell. As the proportional power to be adopted, the former construction of the engines, screw, and paddles, besides Mr Scott Russell, I've had the benefit of the deliberate consideration advice of Mr Field of the firm of Maudslay and Field and of Mr Blake of the firm of Watt and co. I've also written to my friend

Mr FP Smith to whom the public are indebted for the success of the screw, for his advice on the subject. With such assistance I think we may rely upon the certainty of being able to design and execute all that is best in the mechanical and shipbuilding department. In the naval department I have had the opportunity also consulting two gentlemen, Captain Claxton and Captain Robert Ford, who possess special knowledge and experience on the subject. I have had several conferences with all these gentlemen, I explained fully my views, and, with their assistance, settled preliminary some of the principal points of detail.

Brunel's calculations for navigating the Hooghly River. (Brunel Institute)

In February1853 Brunel stepped down from his position with ARMSC and in March he laid out his ideas for the great ship. The ESNC agreed that the first class section should be expanded; the dimensions would be 670 feet long and 80 feet beam. It could carry coal to Diamond Harbour (the highest point accessible up the Hooghly River) from Trincomalee and back. There would be 800 separate cabins larger than those in packet ships, with large saloons capable of accommodating 1,000 or 1,500 first and second-class passengers; and the ship would carry 3,000 tons of cargo. It would make the passage out in about thirty-four and a half days, perhaps thirty-six, steaming at 14 knots on average. For the Australian voyage it would take out 3,000 passengers and a small amount of cargo as considerably more space was needed to carry fuel, since there were then no known coal reserves in Australia.

Tenders were received from Scott Russell, Watt & Co. and Humphreys & Co. for both sets of engines and from Scott Russell for the construction of the hull of the ship. Scott Russell's tender for the ship and paddle engines was accepted and that of James Watt & Co. for the screw engines. Russell signed the final contract in December 1853. In it he agreed that 'all calculations, drawings models and templates which the contractor may prepare shall from time to time be submitted to the engineer for his revision and alteration or approval. Engineer to have entire control over the proceedings and the workmanship.' Brunel had made a rough estimate that the cost of the great ship would be £500,000. Scott Russell's tender for the hull was for £377,200, a figure that has been suggested was a gross underestimate. There had been mention of a sister ship and Russell, additionally, offered to reduce the costs of the whole to £258,000 if he had the contract for that second ship. Finances were to be a major breaking point in the relationship between Brunel and Scott Russell.

Over the next six months matters were not helped by problems within the ESNC. Insufficient capital had been paid up and several shareholders had withdrawn when the mission was changed so new shareholders had to be found. Brunel and one of the directors, Charles Geach, personally acted to gain further financial support by recruiting more shareholders. Geach was an iron founder and man of wealth and was a strong supporter of the project. Such a massive iron ship boded well for the future of his business. He was also a financial supporter of Scott Russell. In previous ship projects, Brunel's role had largely remained that of the consulting engineer, but he had an unusually heavy involvement in the ESNC and became a significant investor.

At one point the directors proposed the appointment of a resident engineer. Heavily resisting the idea, Brunel wrote:

> The fact is that I never embarked in any one thing to which I have so entirely devoted myself, and to which I have devoted so much time thought and labour, on the success of which I have staked so much reputation, and to which I have so largely committed myself and those who are disposed to place faith in me… I cannot act under any supervision, or form part of any system which recognises any other adviser than myself, or any other source of information than mine, on any question connected with the construction or mode of carrying out practically this great project in which I have staked my character. … If any doubt ever arises on these points I must cease to be responsible and cease to act.

The directors backed down.

In November 1854 an article appeared in one of the London newspapers about the great ship and suggested that Brunel was a bystander and that Mr Scott Russell

The *Great Eastern* rising from the stocks. (SS Great Britain Trust)

was the main progenitor. Brunel wrote to Yates, the secretary of the Eastern Steam Navigation Company, furious about his relegation to a mere bystander on the project, which was taking so much of his time, energy and emotion. In his view the article must have been sanctioned by someone within the company. He wrote:

> I have always made it a rule…to have nothing to do with newspaper articles. And for this very reason that I have for so many years shunned public writings, namely, to escape misstatements, I feel compelled on the present occasion to take some stand publicly to correct those erroneous impressions, which must be created by a document having the appearance emanating from ourselves.

He was annoyed that the…

> …spirited merchants of Bristol, who in spite of the strongest condemnation of the plan by the highest authorities, and the ridicule of others, persevered in building and starting the first transatlantic steamer. The circumstances as regards the *Sirius* are coloured so as to be quite incorrect; in the same friendly hand would not thrown ridicule and that by positive full statement upon that which he at the tight same time admits to have been the means of always introducing two of the greatest improvements in steam navigation. A writer wishing success to our enterprise would not have omitted to mention that I had a claim to public confidence on this occasion, for the reason that I was at least the principal advisor in those previously successful attempts. And lastly, I cannot allow it to be stated, apparently on authority, while I have the whole heavy responsibility of its success resting on my shoulders, that I am a mere passive approver of the project of another, which in fact originated solely with me, and has been worked out by me at great cost of labour and thought devoted to it now for not less than three years.

The middle paddle shaft. (*Illustrated London News*)

Getting the paddle shaft onto the ship. (*Illustrated London News*)

The building of the great ship can be divided into distinct stages and the first began in February 1854, lasting until February 1856. It is during this time that Brunel and Scott Russell's relationship descended into acrimony, Brunel as ever wanting detailed information and Scott Russell being rather more general in his use of information. The relationship between Scott Russell and Brunel has been examined by several historians. In Buchanan's words, 'Brunel believed that he had written in his usual insistence that he should be in total control of the engineering side of the project. But Scott Russell, an

Profile view of the *Great Eastern*. (SS Great Britain Trust)

established and successful shipbuilder, expected some latitude to build a ship according to his usual methods, which did not always coincide with Brunel's engineering ideas. This weakness was to prove almost fatal to the project.' The early death of Charles Geach removed a moderating influence just as Robert Bright had been a key influencer in the Great Western Steamship Company. Part of the problem between Brunel and Scott Russell was that their roles were insufficiently clear. Brunel believed that he was in total control of the project while Scott Russell, already a well-established shipbuilder, expected some latitude as perhaps had happened with *Victoria* and *Adelaide*.

Brunel had worked out in detail the internal dimensions for the ship. In March 1853 he asked his old friend Captain Claxton to provide information from other vessels. Claxton wrote to him in March giving comparisons of cabin sizes and dining accommodation from Cunard's *La Plata*, Royal Mail's *Oronoko* [sic] and *Magdalena*, and from the *Great Western* and the *Great Britain*; these two he referred to as 'Old Timers'. In his letter he added one final sentence: 'What a business the *Adelaide* has made of it – where can the Coals have gone to?' This was a reference to the ARMSC ship designed by Brunel, which had just headed out for Australia but was already reported as being short of fuel.

In a personal memorandum in October 1855, Brunel noted his internal designs for the steamship and it is an example of his incredible attention to detail. He considered the location of wash houses, lamp and candle room, spirit stores and inflammable stores, all of which needed to be positioned carefully, either through safety concerns or comfort. As ever, Brunel was keen to try new things and improvements. Protection from fire, which had impacted him so personally on his very first ship, was, he said, '...of considerable importance and I have some hopes the process which has been recently patented by Lt. Jackson may be successfully applied to rendering wood inflammable.' A large icehouse would be required as the vessel would be travelling in tropical waters, plus a smoking room, staircases, ventilation and drainage. Regarding access to the kitchen, he considered 'it would be a good thing to have a railway let into the deck on each side along which a truck can carry dinner et cetera from the kitchen to each saloon.' In this he was not envisaging a steam railway but a simple set of tracks on which a truck could be pushed.

Painting of the *Great Eastern* under construction. (SS Great Britain Trust)

He was concerned rightly about navigation and the problems of compass correction on a vast iron ship and was perhaps sensitive to the problems caused by the stranding of the *Great Britain*. He noted, 'by constant observation to lay down position and course of ship; and correct compasses'. His concerns over this led to correspondence with Professor Airey, the Astronomer Royal, and several new instruments were tested. Airey was an enthusiastic correspondent on the topic, as was Professor Piazza Smyth, the Astronomer Royal for Scotland. Lightning conductors were also debated, and Sir W. Snow Harris was engaged in considering that challenge.

In February 1856, Brunel's battle with Scott Russell came to a head over finances. Russell was stretched to the limit, but Brunel suspended payments, and everything came to a halt. From this point Brunel had greater charge and Scott Russell, in whose yard the ship was being built, was moved to the sidelines. Their personal relationship had broken down completely. Brunel's relationship with the Eastern Steam Navigation Company itself was no easier; in particular, he contrived to upset the secretary to the company, Yates. These were very different working relationships to those Brunel had enjoyed on his previous two ships where he had been able to rely on Claxton, Guppy and Patterson.

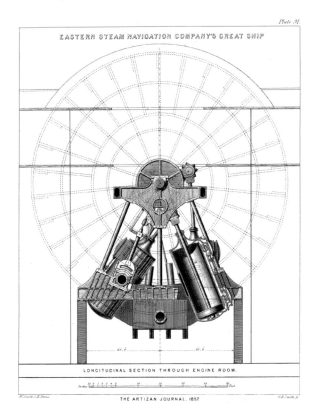

Right: Drawing of the engines.
(SS Great Britain Trust)

Below: Share certificate for
the Great Ship Company.
(SS Great Britain Trust)

Gunboats

Throughout the troubled times of the great ship Brunel was also occupied in war work. During the Crimean War from 1853 to 1856 his brother-in-law and close friend Benjamin Hawes, who had been Under Secretary of State for War and the Colonies until 1851, had involved Brunel in creating a hospital for the Crimea.

With a skill and discipline encouraged by his father, Brunel constantly sketched his ideas. In his many sketchbooks held on behalf the University of Bristol within the SS Great Britain collection, there are endless drawings of ships, from the early drafts of what would become the *Great Eastern* to a small rowing boat, gunboats, sailing vessels,

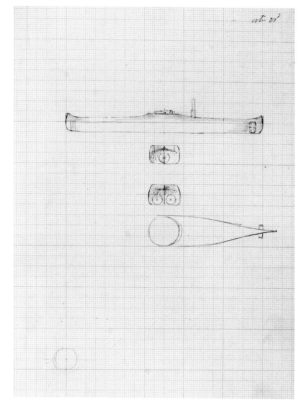

Brunel's sketch of gunboats in 1852. (Brunel Institute)

paddle ships, paddle steamers, cross sections and, even before the *Great Eastern* was launched, a steamship for laying cables across the Atlantic. Brunel designed for business and he designed pleasure boats, including a propeller for a yacht. In one of his sketches, he drew a gunboat with twin screw propellers, giving it a distinctly modern look.

Steam gunboats were a lively discussion in the House of Parliament and in *The Times*. Both Scott Russell and Claxton wrote urging their use in the Baltic. Brunel, who never raised his profile in such way and positively avoided writing to newspapers, drew a variety of gunboats in his sketchbooks and corresponded directly with the Admiralty. As ever a man in a hurry, he found Admiralty departments slow and inefficient. His ideas included floating gun batteries designed to be semi-submersible to present the smallest possible area as a target for enemy fire. These were developed in considerable detail and submitted to the Admiralty, but they remained only intriguing ideas. On 14 January 1856 Brunel wrote to Lord Palmerston urging him to act to 'permit of its being brought into practical operation this season.' It never was.

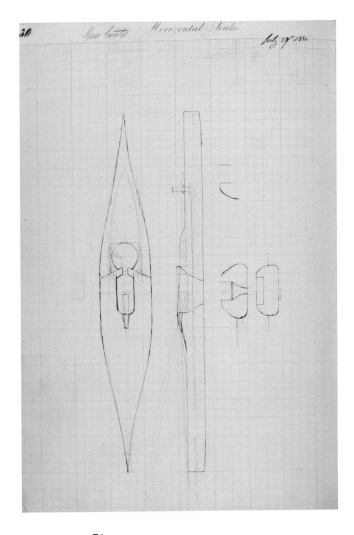

More gunboats, 1854.
(Brunel Institute)

The Great Ship Launch

The great ship moved on apace with a stream of breathless reports in newspapers; it was an ideal subject for the *Illustrated London News*, which had only been in existence since 1842. Another relatively new publication was the satirical magazine *Punch* and when the ship was finally ready for launching in November 1857, *Punch* was there to witness the occasion and, noting the lack of a lunch on offer, wrote a small ditty.

Brunel is a Brick, and Scott Russell's a bean
And their ship is the grandest that ever was seen
And shall still have the aid and the protection of Punch
But today he saw neither a launch nor a lunch

Indeed, the day of the launch had been a major disappointment for everyone. The planned launch was unusual and high risk; due to the length of the ship it was not possible to launch it in the conventional way, stern first into the river. So, Brunel at an early stage had

The Eastern Steam Navigation Company.

MILLWALL, POPLAR, 31st October, 1857.

ADMIT ONE VISITOR

TO THE

LAUNCH OF THE GREAT EASTERN,

Subject to the Regulations of the Company.

It is expected that the Launch will commence on TUESDAY, the 3rd of November, about 11 o'Clock a.m., and will probably last Two days.

The Directors have not been able, until the moment of issuing this card, to determine the period of Launching, and consequently have been unable to provide Refreshment for Visitors.

JOHN YATES, *Secretary.*

N.B.--The only Entrance for Visitors will be at the lower Gate of the Works in the Horseferry Road.

The invitation to the doomed launch. (Clive Richards Collection)

At the projected launch of the great ship, Brunel with his assistant, Jackomb, carrying a chart. (SS Great Britain trust)

decided on a lateral sideways launch and to use iron slipways. Despite careful calculations and the construction of the launch way, the ship did not move. To Brunel's fury this doomed attempt to move the monster was witnessed by large numbers of people who had crowded in to watch the spectacle. The directors had been keen to promote it and to earn some money at last from their expensive project. Tickets were sold and vast numbers turned up but the ship stuck obstinately in the ways and would not be moved.

Punch also had much fun about the sheer scale of the ship. They referred to Captain Harrison and that he would have the 'comparatively unimportant duty of taking care of the vessel.' This was because other captains, they suggested, would be appointed to look after other aspects. There would be a dining captain to handle the large numbers of people dining in the saloons and a flirtation captain to keep the ladies entertained. Further entertainment would be shipboard racing with races taking place around the ship's deck. A *Great Eastern* derby would be a feature of the voyage as a circuit once around the vessel was one third of a mile. A theatre would be built for plays and guest appearances and an English dramatic author would be kept in the hold. They also made fun of that great Victorian institution, the gentlemen's club. *Punch* announced that the clubroom was being arranged and candidates for the *Great Eastern* club had better send in their names. Applicants would be excluded for the following reasons; trade, moustaches, political opinions, whistling, a short pipe, being in the habit of asking questions or a pug nose. Finally, they wrote 'omnibuses will convey humble passengers to various parts of the ship.'

Apart from the unflattering column inches, a side-effect of the unsuccessful launch was that the company and Brunel were now inundated with a stream of suggestions about how the ship could be launched. The helpful writers ranged from curious

Howlett's famous portrait of Brunel in front of the chains. (SS Great Britain Trust)

amateurs, including one young boy, to doctors, clergymen, architects and businessmen and at least one Fellow of the Royal Society. Brunel was supremely uninterested in these suggestions; the only person he felt could advise him was 'my friend Mr R Stephenson'. But he had already decided that it was a simple matter of yet more power, doubling what was needed. He was eventually successful in January 1858 so the final stage, the fitting out of the ship, could commence. Brunel was by now a sick man and illness forced him to take several long absences over the next year, while Scott Russell, who had the contract to fit out the ship, took over. Workmen of all trades invaded the ship including the celebrated firm of Mr Crace who had decorated Brunel's first ship, *Great Western*.

The first paying voyage was tantalisingly near, but the Eastern Steam Navigation Company was unable to recover from its financial woes and a new company was formed which acquired the unfinished ship. The ship was seen as a big insurance risk and it was reported in the *Athlone Sentinel* in August that …

> Few if no policies of insurance have been taken out on the *Great Eastern* at Lloyd's, the underwriters being generally unwilling to transact any business until the completion of the trial trip. Almost the only reason assigned for this course is the possibility of the vessel not getting safely out of the Thames.

Brunel wrote an anxious letter worrying that the ship, which he compared to a half-finished chronometer, was '… capable of becoming a perfect and profitable machine but it was also easy to spend a lot of money and make it a failure.' Finally, the ship was

Above: Members of the public viewing the ship's deck. (*Illustrated London News*)

Right: The final picture of Brunel on board the *Great Eastern*. (SS Great Britain Trust)

Captain Harrison, the first master of the *Great Eastern*. (New York Public Library)

ready and a seriously ill Brunel managed to go on board his ship in September, but there he suffered a stroke which proved fatal a short time later. He lived just long enough to hear of the ship's departure into the English Channel and the tragedy that struck. When just off Dungeness there was a major explosion on board in which five firemen lost their lives and many were injured. In the subsequent inquest into the deaths the coroner and jury were unable to blame any one person for the tragedy. It was by now too late to commence her maiden voyage to New York and she was brought back to Southampton. Here tragedy struck again when the highly regarded first master of the ship, Captain Harrison, drowned in an accident in Southampton Water.

After so many tribulations, not all of their own making, the *Great Eastern* set out on her maiden voyage to New York on 16 June 1860. The first captain so carefully selected, Captain Harrison, had drowned so Captain John Vine Hall was appointed in his place. Just as in the case of Brunel's first ship, the *Great Western*, rumours and bad publicity limited the number of bookings. Despite having accommodation for thousands of passengers there were only thirty-eight paying passengers and eight guests on board. This small number of passengers were on board the most luxurious liner of the day. Again, as in the case of Brunel's previous steamships, no effort had been spared on the interior design. Here again there were innovations, this time with the introduction of a family cabin. An early passenger commented on her comfort as a passenger ship:

> It is hardly possible to say too much in praise of her. She meets all the requirements of the most luxurious hotel, and when the weather drives her inhabitants below they can promenade through her cabins upon long walks, or lounge about superb divans, listening to music that would not discredit the most pretentious concert.

Poster advertising the *Great Eastern*'s maiden voyage to New York, 1860. (New York Public Library)

On the morning of 28 June, the telegraph station at Sandy Hook sent a message that the lights of a large vessel were visible and the *Great Eastern* arrived after a passage of ten days nineteen hours to loom over the New York waterfront. While the crossing time looked good, Captain Vine Hall, who was described as one of the most experienced navigators of the English East India trade, had never crossed the Atlantic before and was deemed to have been somewhat cautious. The company had also taken no chances of embarrassing navigation errors on the approaches to New York harbour by employing Mr Murphy, the senior pilot at New York, and having him join the ship in England.

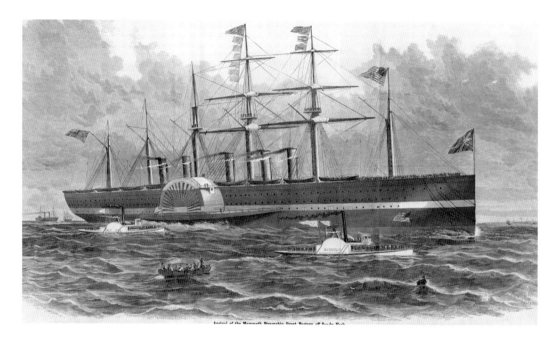

Above: The *Great Eastern* off Sandy Hook on the approach to New York. (New York Public Library)

Left: The *Great Eastern* arriving in New York. (New York Public Library)

On arrival in New York the great ship was warmly welcomed, and thousands visited the ship; in one day there were 18,000 visitors. Other ports were also keen to see the ship for themselves so a two-day excursion to Chesapeake Bay was organised for 31 July, leaving at 3 o'clock in the afternoon. Knowing this was likely to prove very popular it was announced that 'Mattresses will be provided for the men in the various compartments and decks and the staterooms were reserved for the ladies.' Dodsworth's celebrated band would provide the music. 1,500 people bought tickets for the cruise, but the numbers overwhelmed the facilities on board and chaos reigned. Food, drinks and accommodation were in scarce supply and the extra staff hired were completely overwhelmed by the numbers. On return the whole cruise was roundly criticised in the many newspapers for its poor organization. The ship had to be cleaned up for the next trip to Hampton Roads and Annapolis but this time just 105 people travelled on board. Everywhere she visited thousands of people paid their fees to examine the monster and at Baltimore the ship was visited by President Buchanan. The *Great Eastern* returned from Baltimore to New York but with even fewer passengers; just thirty-four people, who paid $20 each. When the ship eventually left on 16 August after so many weeks to head for England via Halifax it was felt she had rather overstayed her welcome. Following a brief visit to Halifax the ship headed for Pembroke with seventy-two passengers.

Daniel Gooch, who was a director of the company, was pleased to leave New York, but he was critical of Captain Hall and his officers in their inability to manage matters on board. On return Hall lost his position, as did the chief engineer, McLennan. In 1861 the ship set off again from Milford Haven with just 100 passengers, heading for the southern states with a new captain, W. B. Thompson, and a new chief engineer, Robertson. The political environment was now difficult as civil war had broken out in the USA. Concerned about the vulnerability of Canada, the British Government decided to send extra troops and, on her return, the *Great Eastern* was pressed into service.

President Buchanan visited the *Great Eastern* in 1860. (New York Public Library)

She carried 2,500 troops and 200 artillery horses to Québec plus wives and families and forty other passengers including William Froude, a colleague of Brunel, and Brunel's son, Henry. The great ship finally fulfilled its destiny as a vast passenger ship, carrying 3,400 people under the command of Captain James Kennedy of the Inman line.

One month later, with 500 passengers, she returned, this time with Captain James Walker at the helm. And so it continued and by the third voyage in 1862 it looked as if most of her problems might be behind her, but under the command of yet another master, Captain Patton, she touched an uncharted reef on the approach to New York in August. For the beleaguered company, yet another accident, yet more repairs and yet more claims, forced them into liquidation. William Hawes, late of the Australian Royal Mail Steam Company, was appointed the official liquidator. The ship was put up for auction in January 1864 in Liverpool. In a familiar story, just as in the case of the *Great Western* and the *Great Britain*, the *Great Eastern* did not reach the reserve price and was withdrawn from sale.

Chaos in the saloon during a gale in September 1861. (*Illustrated London News*).

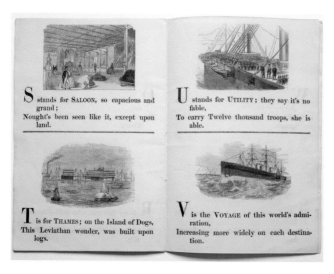

S stands for SALOON, so capacious and grand;
Nought's been seen like it, except upon land.

U stands for UTILITY; they say it's no fable,
To carry Twelve thousand troops, she is able.

T is for THAMES; on the Island of Dogs,
This Leviathan wonder, was built upon logs.

V is the VOYAGE of this world's admiration,
Increasing more widely on each destination.

Page from a Children's Alphabet. (Clive Richards Collection)

The Atlantic Cable

The saviour of the ship was to be a gentleman by the name of Cyrus Field of the Atlantic Cable Company. While cables had been laid across the English Channel and had crossed the Irish Sea, the Atlantic Ocean was the glittering prize. Any ship laying such an extensive cable needed to be large and exceedingly strong and stable, qualities which the *Great Eastern* had in abundance. Indeed, there had been previous suggestions that it was the ideal vehicle. 'Brunel himself it seems first alluded to it in a conversation with Cyrus Field in 1857, and indeed the suggestion had again been made as recently as the company meeting held to elect an official liquidator.' Brunel had, in fact, drawn an outline of a cable ship in one of his sketchbooks.

The failure of the auction persuaded a small group of bondholders with an interest in cable laying, including Daniel Gooch, Cyrus Field and Thomas Brassey Junior, to

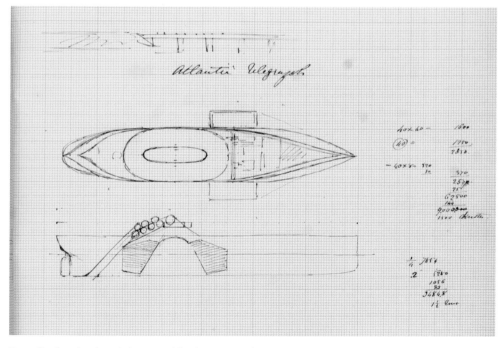

Brunel's sketch of an Atlantic cable ship. (Brunel Institute)

form themselves into a company, the Great Eastern Steamship Company. Gooch was a Brunel protégé, having worked closely with him on the Great Western Railway. They were ready when the ship came up again for auction and they acquired it for a mere £25,000. The ship was then chartered to the newly formed Telegraph Construction Company. One Atlantic cable had been laid and had managed to send 300 messages before breaking. It was a great prize to achieve such speedy communication and there was no shortage of entrepreneurs in a variety of different companies keen to achieve this feat. While the *Great Western* had been warmly welcomed twenty-six years before when she reduced the transmission of news across the Atlantic to a consistent fourteen days, now the possibility of the cable was to reduce it from nine days to mere minutes.

The chief electrician was C. F. Varley who, together with Professor Thomson, was the technical superintendent. The chief engineer was Samuel Canning and the latest captain of the *Great Eastern* was James Anderson, who had been part of the Cunard company. The Admiralty, with a keen eye for the considerable benefits of such rapid communications, offered the possibility of a subsidy in operating the completed cable and provided two warships to assist, *Terror* and *Sphinx*, plus an expert navigator, Captain Moriarty. All of this was arranged at great speed and in October 1864 2,300 miles of cable were carefully coiled in three great water-filled tanks on board the *Great Eastern*. The mechanism for paying out the cable was installed and another mechanism for drawing in the cable was placed at the bow of the ship.

By 14 July 1865 the *Great Eastern* was ready and left the next day to head for Valencia on the south coast of Ireland, where the cable was to begin. On arrival in Ireland the ship attracted its usual intense interest among the locals. The shore cable, 27 miles of which had come on board another steamer, *Caroline*, was attached at the cliffs to the telegraph station and the *Great Eastern* headed off to pick up the end of the shore cable and splice it to the cable on board. Once a signal had been successfully sent to and from

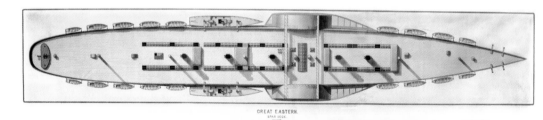

Deck view of the *Great Eastern*. (SS Great Britain Trust)

Side view of the *Great Eastern*. (SS Great Britain Trust)

the shore station, the ship could now begin to head across the Atlantic laying the cable, with signals which were sent every ten or fifteen minutes to and from the telegraph station so that any problems in the cable could quickly be detected. All was well for a few days until, with 745 miles of cable payed out, connection was lost and the cable was laboriously recovered and checked mile by mile until the fault was found. The cable was then cut and re-spliced and relays began again down and up the line. By 2 August, when 1,200 miles of cable had been laid and Newfoundland was just 600 miles away, the cable was accidentally severed. Despite it being 2½ miles down, with incredible effort they almost managed to regain the cable but not quite. So regretfully they were forced to leave the cable, its position marked by a mooring buoy, and leave it where it was and head for home.

With great determination of which Brunel himself would have been justly proud, Cyrus Field and Daniel Gooch established another company called the Anglo-American to make another attempt at laying the cable. The lessons that had been learnt from the previous attempt improved both the apparatus on board and the cable itself. In July 1866 the *Great Eastern* headed off for the second attempt, this time accompanied by three Admiralty vessels, the *Terrible*, *Medway* and *Albany*. Daniel Gooch accompanied the venture together with Captain Anderson and with Samuel Canning as the engineer in charge.

She made steady progress after leaving Valencia on 13 July and emerged from the fog into Heart's Content Bay on 27 July, having achieved the laying of the cable in just fourteen days.

LANDING OF THE SHORE END AT TRINITY BAY, AUG. 4TH 1858.

Taking the cable ashore during the first short-lived attempt. (New York Public Library)

TRINITY BAY AND HEART'S CONTENT, NEWFOUNDLAND, THE LOCALITY WHERE THE WESTERN END OF THE ATLANTIC CABLE WAS TO HAVE BEEN LAID.—From a Sketch by our Special Artist, Mr. J. Becker.

Heart's Content Bay. (New York Public Library)

Before her return voyage, the vast ship needed to refuel, and as the bay was simply a small village with a few houses, mainly belonging to fishermen, five ships had sailed from Cardiff with coal to await the arrival of the *Great Eastern*. Also waiting for them was the rest of the 1865 cable, 600 miles of which had come out in the *Medway*. The naval captain, Captain Moriarty, went out in the *Albany* to find the spot at which the old cable had broken. By the time the *Great Eastern* arrived at the position there was a neat line of buoys marking the end of the cable. By drifting across the line with a three-mile long grappling hook the cable eventually was attached and after considerable efforts the old cable was linked again, and transmission confirmed. The *Great Eastern* headed once more for Heart's Content Bay and now to the delight of both Cyrus Field and Daniel Gooch there was not just one but two working telegraph cables spanning the Atlantic. It was a moment to cherish. The great achievement was recognised, and the company made a record dividend of 70 per cent on the *Great Eastern* steamship shares. Official recognition came with several knighthoods, including one for James Anderson, the master of the vessel.

The Paris Exhibition of 1867 supplied the next role for the ship when a French company chartered the vessel to act as a passenger ship between New York and Brest. It was refitted once again as a passenger vessel. However, it experienced severe storms which delayed the ship on its way to New York. Among the passengers was Jules Verne, the writer, who had booked for a round trip, and the ship returned to Brest with just 180 passengers. The French company was in debt and the ship was at risk of being seized. The crew was unpaid and sued via the Court of Admiralty. Once matters were resolved, albeit not well, the Great Eastern Steamship Company was unable to pay any dividends. However, her cable laying days were not over as the Société du Cable Transatlantique Francais was the next to charter the *Great Eastern*. Under the command of Captain Halpin, who had been first mate on the previous cable laying, in 1869 she successfully laid a cable between Newfoundland and Brest, much to the delight of P. J. Reuter, a naturalized Briton who had developed a news and intelligence agency.

Municipal Dinner

TO

CYRUS W. FIELD, ESQ.,

AND OFFICERS OF

H. B. M. STEAMSHIP GORGON,

AND

U. S. STEAM FRIGATE NIAGARA,

IN COMMEMORATION OF THE

LAYING OF THE ATLANTIC CABLE,

Metropolitan Hotel, Sept. 2, 1858.

REGULAR TOASTS.

1,—THE PRESIDENT OF THE UNITED STATES.

2,—THE QUEEN OF GREAT BRITAIN AND IRELAND :

3,—THE GOVERNMENT & PEOPLE OF GREAT BRITAIN & IRELAND :
Joined to us in the Court of Neptune, may the nuptials never be put asunder.

4,—OUR SISTER STATES AND SISTER CITIES :
New York greets them ; and trusts that the cord which binds the New World with the Old, welds more firmly the links of the Union.

5,—CYRUS W. FIELD :
To his exertions, energy, courage and perseverance, are we indebted for the Ocean Telegraph ; We claim, but Immortality own's him.

6,—THE NAVIES OF GREAT BRITAIN AND THE UNITED STATES :
Met and joined in a noble work of Peace—may they never be separated or meet in Strife.

7,—THE ENGINEERS AND ELECTRICIANS :
Who have done their work so well. The praises of both Hemispheres shall be their reward.

8,—THE STATE OF NEW YORK :
May her history always illustrate her motto—Excelsior !

9,—OUR CITY OF MANHATTAN :
Foremost of America, now placed side by side with the chief Cities of Europe ; while we strive to surpass we will be friends as well as rivals.

10,—THE NEW YORK, NEWFOUNDLAND AND LONDON COMPANY
Which commenced and planned, and the ATLANTIC TELEGRAPH COMPANY which completed the work of linking two Continents together beneath the Sea. They have done more for the Civilization and Peace of the World than any other companies which ever existed.

11,—THE ARTS OF PEACE :
Now crowned with immortal lustre, and for once at least covered with greater renown than ever were the Arts of War.

12,—THE PRESS :
To which the Telegraph is both minister and instrument ; may its usefulnsss be always equal to its powers.

13,—WOMAN :
At whose feet we lay all our triumphs ; to her we owe the happiness of life and the consolations of home. God bless her !

Dinner in New York for Cyrus Field. (New York Public Library)

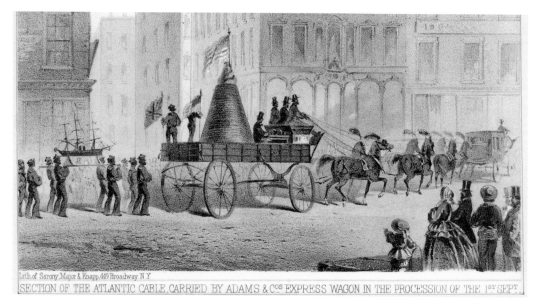

Lith.of Sarony,Major & Knapp,449 Broadway N.Y.

SECTION OF THE ATLANTIC CABLE, CARRIED BY ADAMS & Cᵒˢ EXPRESS WAGON IN THE PROCESSION OF THE 1ˢᵀ SEPT.

Above: Celebrating the cable in New York. (New York Public Library)

Below left: Jules Verne, French author. (New York Public Library)

Below right: Celebration for Captain Halpin on laying the French cable. (New York Public Library)

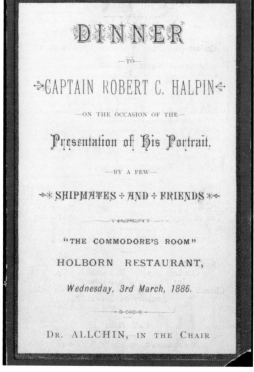

DINNER

—TO—

CAPTAIN ROBERT C. HALPIN

—ON THE OCCASION OF THE—

Presentation of His Portrait,

—BY A FEW—

SHIPMATES AND FRIENDS

"THE COMMODORE'S ROOM"

HOLBORN RESTAURANT,

Wednesday, 3rd March, 1886.

Dr. ALLCHIN, IN THE CHAIR

At the end of that year she fulfilled one of her early destinies. The original plan had been for the great ship to run to India as well as Australia and Brunel had designed the vast vessel with a relatively shallow draft so that she could proceed safely up the Hooghly River to Calcutta. In January 1870 she laid a cable between Bombay and Suez. For two years after this the ship languished idle in the Mersey River but her cable laying days were not over yet. In May 1873 she laid her fifth cable, her fourth across the Atlantic, and in the summer of 1874 she made yet another cable laying voyage, this time in the reverse direction from Heart's Content Bay to Valencia. During this period, she was financially successful and provided good dividends to shareholders; however the company's financial success was dented by a fraud committed by its secretary.

The ship returned to Milford Haven to await the next stage. Conversion back to a passenger ship was going to be incredibly expensive for what was now an ageing vessel and in 1881 she was advertised for sale but did not reach the minimum price of £75,000. There were various schemes to charter the ship, few of which came to anything, and so in 1885 the ship was yet again up for sale and her most colourful period of service commenced. The ship was sold to a gentleman called Edward de Mattos who chartered the ship to an emporium and early department store in Liverpool called Lewis's. The ship made her way from Milford Haven to Liverpool and the ship was covered in advertising signs for Lewis's. She was moored in the Mersey and became a great tourist sight. She was eventually purchased in July 1887 for £26,000 by a Mr Walmsley who converted the ship into a 'pleasure resort', with music halls, amusement arcades and even trapeze artists swinging between the second and third masts. By the end of that summer the great ship was on the move again. From Liverpool she made a trip up to Greenock and anchored in the Clyde. The pleasure business was not paying as well had been hoped and eventually the ship was sold to Bath & Company, Liverpool and London metal brokers, for £16,500. And so, in August 1888 she made her last voyage from the Clyde down to the Mersey, where she was grounded. Her fittings were sold the following year and the ship was gradually dismantled over two years until there was nothing left.

The *Great Eastern* moored in the Mersey. (SS Great Britain Trust)

The *Great Britain* in the Twentieth Century

By the end of the nineteenth century, all of Brunel's big ships had long since ceased operating. The *Great Western* was broken up in 1857, the *Great Eastern* in 1888 while the *Adelaide* and *Victoria* were both sold in 1860 and their fate is unknown. The *Great*

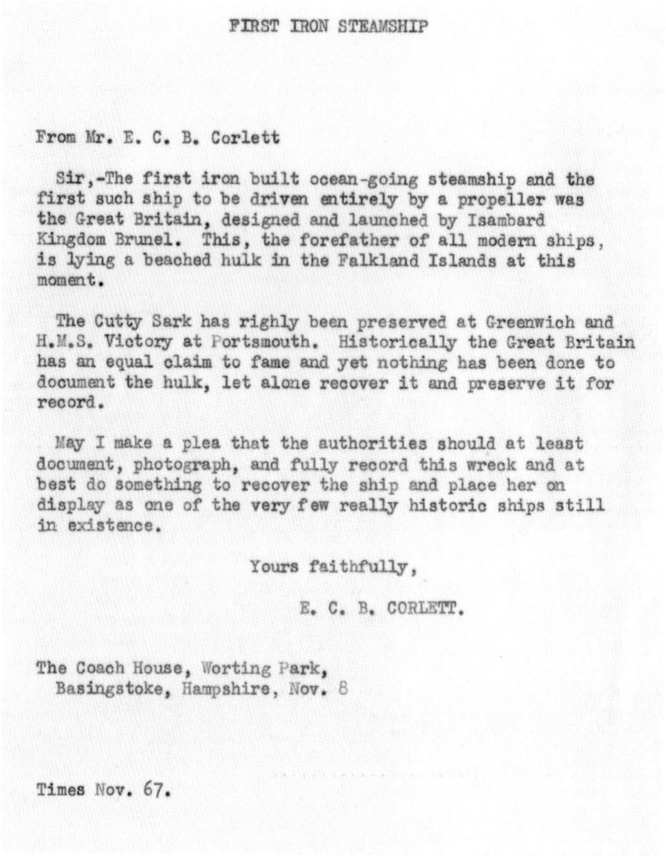

```
                    FIRST IRON STEAMSHIP


  From Mr. E. C. B. Corlett

     Sir,-The first iron built ocean-going steamship and the
  first such ship to be driven entirely by a propeller was
  the Great Britain, designed and launched by Isambard
  Kingdom Brunel. This, the forefather of all modern ships,
  is lying a beached hulk in the Falkland Islands at this
  moment.

     The Cutty Sark has righly been preserved at Greenwich and
  H.M.S. Victory at Portsmouth. Historically the Great Britain
  has an equal claim to fame and yet nothing has been done to
  document the hulk, let alone recover it and preserve it for
  record.

     May I make a plea that the authorities should at least
  document, photograph, and fully record this wreck and at
  best do something to recover the ship and place her on
  display as one of the very few really historic ships still
  in existence.

                    Yours faithfully,

                       E. C. B. CORLETT.


  The Coach House, Worting Park,
    Basingstoke, Hampshire, Nov. 8



  Times Nov. 67.
```

The draft of Corlett's letter to *The Times*. (SS Great Britain Trust)

Britain remained by an accident of fate, abandoned and forgotten. Her register was closed on 18 December 1886 and on 8 November 1887 Anthony Gibbs sold the *Great Britain* to the Falklands Island Company. She became a hulk in the Falklands, a floating store ship for wool and coal, and remained there at anchor. In 1936 there was an abortive attempt by the Governor of the Falklands, Sir Herbert Henniker Heaton, to have the ship preserved and returned to England but with no success. Even her days as a hulk were over and she was left in Sparrow Cove, 3½ miles from Stanley, where she was run aground and scuttled.

It would not be until 1967 that another attempt to retrieve the ship began in all seriousness when Ewan Corlett wrote to *The Times* newspaper in November, urging action to save the ship for the nation and setting it alongside the *Cutty Sark* and HMS *Victory* as national treasures. Corlett's attention had been drawn to the ship by the then director of the San Francisco Maritime Museum, Karl Kortum. Corlett's letter triggered a very positive response which led to surveys of the ship and the beginning of a major task to raise funds for its return to Bristol. With great determination and the support of many people such as Jack Hayward and Richard Gould Adams, finances were raised. With practical support from the government and the Royal Navy the ship was floated onto a raft and on Wednesday 6 May she began her last great passage across the Atlantic. She arrived at Avonmouth on 24 June. On 19 July 1970 the *Great Britain* returned to her original dock that she had left exactly 127 years ago. She was greeted by great crowds including His Royal Highness Prince Philip, Duke of Edinburgh, in the same way that his predecessor Prince Albert had waved out the ship in the last century. It was an incredible feat of determination by a small group of enthusiasts to bring this amazing ship back to its home. The fact that we are still able to visit the ship and admire her is a fitting tribute to her workmanship and design overseen by Isambard Kingdom Brunel.

The *Great Britain* being escorted on her transatlantic return passage. (SS Great Britain Trust)

Towing the *Great Britain* round the Avon's Horseshoe Bend in 1970. (SS Great Britain Trust)

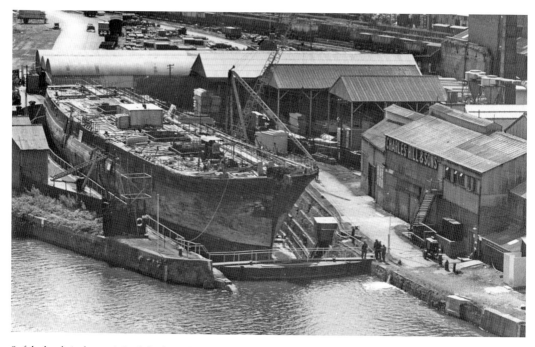

Safely back in her original dock. (SS Great Britain Trust)

Brunel's Impact on the Maritime World

Brunel had a very personal connection to his great ships, typified by his fury over the stranding of the *Great Britain*. During the building of the *Great Eastern*, Brunel travelled to Vauxhall for one final look at his first ship, the *Great Western*, which was then being broken up, and his passionate single-minded dedication to the *Great Eastern* was evident. This giant ship was to remain the largest in the world throughout the nineteenth century until 1899 when the White Star Line launched the *Oceanic* from Belfast. These ships were not Brunel's only maritime ventures and his influence on the world of naval architecture spread far and wide from industrial craft, leisure boats and yachts to naval gunboats and warships.

His first great Atlantic liner, the *Great Western*, influenced wooden paddle ships for some years including the Cunard Line, Royal Mail Steam Packet Line and the navy's

The *Great Britain* today. (SS Great Britain Trust)

packet ships. The *Great Britain* led the way in propeller technology and the use of iron, promoting the use of the latter even when it was stranded. A sister ship, *Royal Charter*, was commissioned by Gibbs Bright and, when the original shipbuilder went bankrupt, was completed by Patterson, who had built both of Brunel's Bristol ships. That the *Great Britain* has survived through lengthy service across the Atlantic, the Indian Ocean, around Cape Horn and abandonment for so many years in the Falklands is a testament to its designer. His work for the navy as a consultant on trials of HMS *Polyphemus* and the design of HMS *Rattler* ensured the commissioning of propeller-driven warships. The *Great Britain* influenced the first iron warship, *Warrior*. All of Brunel's great ships led the way in design both externally and internally and the provision of passenger facilities. They enabled fast communications and connected the world, notably by the cables laid by the *Great Eastern*. They all provided service to the British government carrying mail and vital despatches around the world and took troops and horses to the Crimea, Canada and India. In the Crimea not just two of Brunel's ships but four were in service as the *Victoria* and *Adelaide* joined the *Great Western* and *Great Britain* in the Mediterranean and Black Seas. The magnificent *Great Britain* remains today as an inspiring monument to a truly exciting time in both naval architecture and oceanic travel, and to a world class engineer.

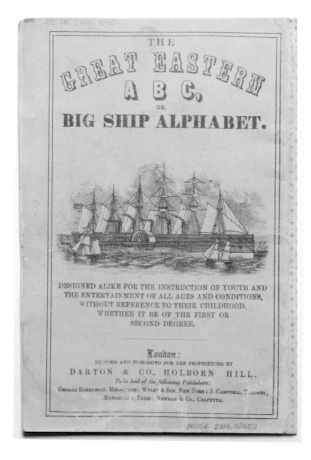

Above: The *Great Western*'s ship's bell. (SS Great Britain Trust)

Left: The children's alphabet of the *Great Eastern*. (Clive Richards Collection)

Time Line

1817	Brunel witnesses his father's steamboat trials
1820	Brunel writes home about making model boats
1827	Brunel writes of ambition to build boats and shipyard
1832	Designs a drag boat for Bristol's floating dock
1835–37	Consulting engineer on *Great Western*
1837	Launch of *Great Western*
1838–43	Consulting engineer on *Great Britain*
1840	Propeller sketches
1843	HMS *Rattler*, Royal Navy
1843	Launch of *Great Britain*
1845	Brunel drawing of ship with two paddlewheels
1846	*Great Britain* stranded in Dundrum Bay
1847	*Great Britain* salvaged
1850	Sale of *Great Britain* to Gibbs Bright
1851–53	Consulting engineer for Australian Royal Mail Steam Ship Company
1852	*Adelaide* and *Victoria* for Australian Royal Mail Steam Ship Company
1852	Drawing of twin-screw gun boat
1852	Drawings of *Wave Queen*, *Victoria*, *Great Britain* and unnamed vessel
1853–58	The Great Ship project commences for Eastern Steam Navigation Company
1854	Drawing of gun boat with turret
1854	Letter to Brunel from Claxton about *Great Britain*
1854	Correspondence about gunboats with Admiralty
1857	Brunel's final look as *Great Western* is broken up
1858	Launch of *Great Eastern*
1858	Death of Brunel

Great Britain lit up at night. (SS Great Britain Trust)

Selected Further Reading

Bagust, Harold, *The Greater Genius? A Biography of Sir Marc Isambard Brunel* (Hersham: Ian Allen Publishing, 2006).

Brunel, Isambard, *The Life of Isambard Kingdom Brunel, Civil Engineer*. Originally published 1870 (Stroud: Nonsuch Publishing, 2006).

Buchanan, R. Angus, *Brunel: The Life and Times of Isambard Kingdom Brunel* (London: Hambledon and London, 2002).

Corlett, Ewan, *The Iron Ship: The Story of Brunel's SS Great Britain*. Originally published 1975 (Bristol: SS Great Britain, 2012).

Doe, Helen, *The First Atlantic Liner, Brunel's Great Western Steamship* (Amberley Books, 2017).

Emmerson, George S., *SS Great Eastern: The Greatest Iron Ship* (David & Charles, 1980).

Greenhill, Basil (ed), *The Advent of Steam* (Conway Maritime Press, 1992)

Griffiths, Denis, Andrew Lambert, and Fred Walker, *Brunel's Ships* (London: Chatham Publishing, 1999).

Marsden, Ben, and Crosbie Smith, *Engineering Empires: A Cultural History of Technology in Nineteenth-Century Britain* (London: Palgrave, 2005).

Rolt, L. T. C., *Isambard Kingdom Brunel*. Originally published 1957 (London: Penguin, 1976).

Vaughan, Adrian, *Brunel: An Engineering Biography* (Ian Allan, 2006).

Young, Chris, *The Incredible Journey: The SS Great Britain Story 1970–2010* (Bristol: SS Great Britain, 2010)

Image Acknowledgements

A large number of the illustrations in this book are held at the SS Great Britain Trust in Bristol. They hold the nationally recognised collection of items relating to Isambard Kingdom Brunel. Within this there are two distinct collections:

Clive Richards Collection: Accepted under the Cultural Gifts Scheme by HM Government from Clive Richards OBE DL and allocated to the SS Great Britain Trust, 2017.

Brunel Institute: A collaboration of the SS Great Britain Trust and the University of Bristol.